DEPARTMENT OF THE ENVIRONMENT

Waste Management Paper N

Landfilling Wastes

A Technical Memorandum on the Legislation, Assessment and Design, Development, Operation, Restoration and Disposal of Difficult Wastes to Landfill including the Control of Landfill Gas, Economics, a Bibliography and Glossary of Terms

London: HMSO

ISBN 0 11 751891 3

HMSO publications are available from:

HMSO Publications Centre
(Mail, fax and telephone orders only)
PO Box 276, London, SW8 5DT
Telephone orders 071-873 9090
General enquiries 071-873 0011
(queuing system in operation for both numbers)
Fax orders 071-873 8200

HMSO Bookshops
49 High Holborn, London, WC1V 6HB
(counter service only)
071-873 0011 Fax 071-873 8200
258 Broad Street, Birmingham, B1 2HE
021-643 3740 Fax 021-643 6510
Southey House, 33 Wine Street, Bristol, BS1 2BQ
0272 264306 Fax 0272 294515
9-21 Princess Street, Manchester, M60 8AS
061-834 7201 Fax 061-833 0634
16 Arthur Street, Belfast, BT1 4GD
0232 238451 Fax 0232 235401
71 Lothian Road, Edinburgh, EH3 9AZ
031-228 4181 Fax 031-229 2734

HMSO's Accredited Agents
(see Yellow Pages)

and through good booksellers

DEPARTMENT OF THE ENVIRONMENT
2 MARSHAM STREET LONDON SW1P 3EB
01-212 3434

My ref:

Your ref:

The Control of Pollution Act introduced a system of local authority control over waste management based on the licensing of individual sites. This is backed up by the advice and guidance on appropriate disposal technologies provided in Waste Management Papers prepared by my Department.

At present over 5000 licences are in operation, of which about 80% are for waste disposal activities involving landfill. This Waste Management Paper takes particular account of the comments made on hazardous waste disposal by the House of Lords Select Committee on Science and Technology (the Gregson Committee), also the Hazardous Waste Inspectorate in their first report and the Eleventh Report of the Royal Commission on Environmental Pollution. The guidance in it should help to achieve high standards for waste disposal to landfill throughout the country.

Its central theme is the need to plan in a professional way all stages of disposal operations. This makes it easier to monitor what is done and to ensure that the requirements of the licensing authorities are met.

Contributions to the informal report of the Landfill Practices Review Group, upon which this Waste Management Paper is based, came from a wide range of specialists. I should like to thank members of the group for making available their knowledge experience and time.

ANGELA RUMBOLD

CONTENTS

CHAPTER 1. INTRODUCTION

Background to the report

1.1 Developments since 1974 have confirmed that properly managed landfill (the disposal of wastes onto land) is the main disposal route for wastes in the UK. Of the 100 million or so tonnes of controlled wastes that arise each year, over 90 per cent go directly to landfill. The remaining wastes are treated in some way, the residuals from treatment then almost always going to landfill. However, the disposal of wastes to land has the potential to cause severe environmental pollution and pose risks to human health. Accordingly, both a legislative framework and a body of practice and experience have been built up to ensure that waste disposal is safe and acceptable. In only a few instances in recent years has waste disposal led to significant environmental pollution.

1.2 There can, however, be no room for complacency. If landfill is to remain pre-eminent among disposal options, consistently high standards are essential and these standards are subject to change. The requirements that waste disposal technology must meet are continually changing. As society has become more affluent the quantities of waste to be deposited have increased and waste types have changed and increased in number. Similarly as the manufacturing base of the economy diminishes and the high technology and service sectors expand, so the nature of wastes from industry inevitably changes. In addition, new knowledge gained from research and field experience can be expected to lead to changes in our appreciation of the standards that need to be observed. Waste disposers must be alive to these changes and modify their practices or adopt new ones as required.

1.3 It is the view of the Department of the Environment that the time is ripe for a review of the existing guidance on landfill practice, to take note of changing circumstances and to consolidate the knowledge gained from research with the practical experience gained in the field. In reaching this view the Department has taken note of the opinions of others, and particularly of the report on Hazardous Waste Disposal by the House of Lords Select Committee on Science and Technology (The Gregson Committee). A number of the recommendations in that report relate to the need to ensure that only the highest standards are adopted at landfills.

1.4 As part of its response to the Gregson Committee's recommendations, the Department set up the Landfill Practices Review Group in 1982. The Group was given the task of providing guidance and advice on the procedures and practices that should be employed to ensure the continued acceptability of landfill as a disposal route for wastes. This Waste Management Paper is based on the Group's findings.

Structure of the waste management paper

1.5 In the main the paper covers the various aspects of landfilling, in the sequence in which they would arise, from the planning of a new site to its eventual restoration. It will be apparent however that various aspects of a complete landfill design and operation will need to be considered at many and in some cases, at all stages. The paper attempts to avoid unnecessary duplication by dealing fully with each topic where this appears most relevant and by providing appropriate cross-references elsewhere. The following paragraphs give a brief indication as to the content of subsequent Chapters in the paper.

Legislative framework (Chapter 2)

1.6 The main phases of landfill waste disposal are: finding a suitable site; preparing it to accept wastes; operating it using best practices; and finally its restoration and aftercare. In the UK this progression takes place within a framework of legislation which aims to control all stages of the project and thereby ensures that protection of both public health and the environment. Good landfill practice calls for an understanding of, and compliance with, this legislation; the landfill operator must be aware of the duties which legislation imposes and take these duties into account at each stage of the project. A summary of the relevant legislation is given in Chapter 2.

Assessment and design (Chapter 3)

1.7 Good landfill practice starts with the selection of a suitable site and, with planning and design, to prepare it for waste disposal operations. Detailed site investigation, assessment and design have been identified as the cornerstones of good practice. Chapter 3 considers these and stresses the need for an ordered programme of investigation and appraisal, followed by incorporation of the results obtained and conclusions reached into plans for the preparation, operation, reclamation, and restoration of the landfill to beneficial after use. Planning in advance is fundamental to the development and operation of a modern landfill.

Preparation (Chapter 4)

1.8 No site is likely to prove suitable for landfilling wastes without some preparation. While the amount and type of preparation required is specific to the individual site, two main forms of preparatory works have been identified and these are discussed in Chapter 4. First, work related to the infrastructure of the site such as roads, facilities, fences and the like, is common to all landfills. Second, work related to engineering the site to be able to accept the range of wastes planned. These latter works concern measures to control pollution and are specific to the site and the wastes to be deposited. While these works may also be appropriate at sites taking difficult wastes, special care needs to be taken to ensure that the facilities provided and materials used are compatible with such wastes. These special considerations are discussed in Chapter 7 (see also paragraph 1.11 below).

Operation (Chapter 5)

1.9 Once preparations are complete the site is ready to accept wastes. However, well in advance of this, detailed consideration should have been given to the operational practices to be employed at the site and the necessary safety measures to be adopted. Good practice is essential not only to maximise the life of what might be a scarce facility, but also to ensure that the environment and any nearby communities are suitably safeguarded from pollution. Chapter 5 reviews the factors that need to be evaluated when drawing up the operational plan for a landfill, and the various practices to be used during its operation. Here again the types and quantities of wastes deposited will be a major consideration. Special practices that may need to be adopted for some difficult wastes are discussed in Chapter 7 (see also paragraph 1.11 below).

Restoration (Chapter 6)

1.10 The continued acceptance of waste disposal by landfill relies on restoration to agreed standards and planned afteruses. Chapter 6 provides guidance and advice on how restoration should be planned and accomplished, and post-closure management when landfill sites are restored. Although the principles outlined here will apply to all landfills, special considerations may apply to sites that have taken difficult wastes; these are discussed in Chapter 7 (see also paragraph 1.11 below).

The disposal of difficult wastes (Chapter 7)

1.11 Generally the considerations set out in Chapters 2-6 apply to the landfilling of all kinds of wastes. However, a number of special factors arise in regard to the design, preparation, operation and restoration of landfill sites taking difficult wastes and these are discussed in Chapter 7.

Membership of the Landfill Practices Review Group (Annex 1)

1.12 The membership of the Landfill Practices Review Group was drawn from many of the authorities and organisations involved in waste management. In order to cover the wide range of topics on which advice was needed, two sub-groups were formed. One sub-group concerned itself with the general principles behind landfill disposal; the other considered the practices associated with the disposal of industrial and, in particular, difficult wastes. The Chairmen of the Review Group and its Sub-Groups were drawn from organisations that are intimately involved in setting standards for waste disposal. Membership of the Group and its sub-groups is listed in Annex 1.

Landfill Gas Standards Committee (Annex 2)

1.13 A topic which has been identified as being important is the exploitation of landfill gas as a source of low-grade energy. There is a need to consider how this gas might be utilised safely and effectively. As an important step, the report of the Department of the Environment's Landfill Gas Standards Committee is reproduced as Annex 2.

Bibliography (Annex 3)

1.14 In addition to details of official publications on waste disposal Annex 3 provides a list of literature references from which both general and specific information may be obtained.

The economics of waste disposal (Annex 4)

1.15 No review of the landfilling of wastes can be considered to be complete without a knowledge of the cost involved for the various waste disposal options available. Annex 4 provides indicative costs of the range of operations necessary to establish, operate and provide aftercare for a modern landfill site using best current practice. At the same time it also provides a check list of the various considerations necessary to achieve this.

Glossary (Annex 5)

1.16 Waste disposal brings together a range of professions each with its own understanding of the technical terms it uses. A glossary of many of these terms is included.

CHAPTER 2: LEGAL CONSIDERATIONS

Introduction

2.1 The selection, preparation, operation and restoration of landfill sites in England and Wales, and with certain differences in Scotland,* is principally subject to control under two Acts of Parliament. First, the Town and Country Planning Act 1971 (Scotland 1972), as subsequently amended, provides the planning background and control to which all development of land is subject. The second is the Control of Pollution Act 1974, Part I of which is concerned with waste collection and disposal and, in particular requires all disposal sites for controlled wastes to be licensed. There are a number of EC Directives that relate to landfill operations, which have been implemented through the Control of Pollution Act 1974 and its relevant regulations. The Health and Safety at Work etc Act 1974 is also relevant. A summary of the main legislative provisions regulating landfill operations is given in Table 2.1 (see also paragraph 2.79).

2.2 It should be noted that changes in legislation affecting waste disposal have resulted from the abolition of the Greater London Council (GLC) and the Metropolitan County Councils in 1986. Also, draft proposals for amendment of the Control of Pollution Act (Licensing of Waste Disposal) Regulations 1976 and 1977 are being considered, as are other legislative proposals to improve enforcement provisions including aftercare.

2.3 While planning and licensing may rightly be seen as separate stages through which any proposed landfill development must pass, the close interactions between them must be borne in mind. Except for certain permitted developments (see paragraph 2.9) planning permission must be obtained before a waste disposal licence can be issued. On the other hand, it is clearly desirable to establish before planning permission is given that, with any necessary conditions, a landfill development is likely to qualify for licensing. Many of the considerations involved will be common to both planning and licensing decisions and indeed, increasingly, the two stages proceed more or less in parallel. In this chapter the legislative provisions relating to the planning and licensing of landfill developments, and the interactions between them, are described. The requirements of the Health and Safety at Work etc Act 1974 and other relevant legislative are then set out. Finally, other legislation that has some bearing in waste disposal is listed.

Planning legislation and development control

The Town and Country Planning Act

2.4 The 1971 Act (1972 Scotland) makes provision for the control of the use of land for waste disposal through:

Table 2.1 Main legislative provisions for the control of waste disposal by landfill

NATIONAL LEGISLATION

Planning

Town and Country Planning Act 1971 as amended

Town and Country Planning (Scotland) Act 1972 as amended

Local Government Act 1972 and 1985

Local Government (Scotland) Act 1973

Town and Country Planning General Development Order 1977 as amended

Town and Country Planning General Development Scotland Order 1981

Local Government and Planning (Scotland) Act 1982

Pollution

Public Health Act 1936

Control of Pollution Act 1974

Refuse Disposal (Amenity) Act 1978

Civic Government (Scotland) Act 1982

Safety

Health and Safety at Work etc Act 1974

EC DIRECTIVES IMPLEMENTED BY NATIONAL LEGISLATION

Directive on the Disposal of Waste Oils (75/439/EEC)

Directive on Waste (75/442/EEC)

Directive on the Disposal of Polychlorinated Biphenyls and Polychorinated Terphenyls (76/403/EEC)

Directive on Pollution Caused by Certain Dangerous Substances Discharged into the Aquatic Environment (76/464/EEC)

Directive on Toxic and Dangerous Waste (78/319/EEC)

Directive on the Protection of Groundwater Against Pollution Caused by Certain Dangerous Substances (80/68/EEC)

---

* In a number of respects the legislative provisions in Northern Ireland differ significantly from those that apply in England and Wales, and in Scotland. In order to simplify presentation the particular position in the Province has not been covered in this chapter or in other relevant areas of the report. However, it will be apparent that the principles and practice of good landfill development and management, as set out in this report, apply in Northern Ireland as elsewhere in the UK.

(a) structure plans which set out the policies and general proposals for the development and other use of land;

(b) local plans which relate the policies of the structure plan to precise areas of land, and

(c) the grant or refusal of planning permission.

There are differences, as between England, Scotland and Wales, in the levels at which these responsibilities are exercised by local authorities. The position is summarised in Table 2.4 which also shows where responsibility for mineral planning and waste disposal rests. Mineral planning has a special relevance to waste disposal since landfill operations are often located in mineral extraction sites. As noted above, sensible control of waste disposal to land also requires close co-operation within approving councils between planners and waste disposal officers. Indeed, it was recognition of this need that led to planning functions for waste disposal sites in England being made the responsibility of the County Planning Authority (Town and Country Planning (Prescription of County Matters) Regulations 1980: the provisions in the Local Government Act 1985 mean that planning and licensing responsibilities for waste disposal are both at Borough/District Council level following abolition of the Metropolitan Counties and the Greater London Council except in the case of licensing where the Secretary of State has established a Statutory Authority under Section 10 of the Act.

Structure and local plans

2.5 The present development plan system is set out in the 1971 Act (Scotland 1972) (as amended) and secondary legislation. Structure and local plans are replacing the old development plans prepared under the provisions of earlier planning legislation. The system comprises two levels of planning. The structure plan, which is prepared by the strategic planning authority, deals with issues of structural importance for a county (region) or part of a county (region). It consists of a written statement and a key diagram which is not on a map base. This may be followed by local plans which carry forward in detail the general proposals of the structure plan, and relate them to precisely defined areas of land.

2.6 Structure and local plans provide an opportunity for the requirements and constraints of waste disposal strategies to be considered in the overall context of the development and other use of land. In particular, they allow an opportunity to examine the relationship between the waste disposal strategy and policies for mineral working.

2.7 The structure plan will, where appropriate, set out the broad criteria to be applied in identifying waste disposal sites. It may include a general indication of the areas in which additional provision is to be made having regard to the objectives of the waste disposal plan required by Section 2 of the Control of Pollution Act 1974 (see paragraph 2.45 et seq). If more detailed site-specific guidance is required it will be for the planning authority, in consultation with other authorities, to decide whether the particular needs of their area would be better served by a subject local plan covering a wide area or by the incorporation of waste disposal proposals in more general local plans.

Planning control

2.8 Planning permission is necessary for the use of land for the disposal of waste and the carrying out of ancillary development. In granting planning permission the planning authority may, and usually will, impose conditions to minimise damage to amenity or to protect other interests, and to secure suitable and effective restoration of the land.

Table 2.4 Local Authority responsibilities for planning and waste disposal in England, Scotland and Wales

	Structure/ strategic planning	Local plans	Determination of planning applications for landfill	Minerals planning	Waste disposal
England[1]	County	County/ District	County[2]	County	County
Wales	County	District	District	County	District
Scotland	Regional/ Islands	District[3]/ Islands	District[3]/ Islands	District[3]/ Islands	District/ Islands

[1] For England, the references to county council in the Table include the GLC and the Metropolitan authorities. New arrangements have been introduced as a result of the abolition of these authorities in 1986 (see paragraph 2.2).

[2] In England planning applications for landfill development should be made to the appropriate district council (or borough or city in London) but are referred to the county for determination.

[3] In Scotland the responsibilities are exercised by the district councils except in the Borders, Highland and Dumfries and Galloway Regions, where the regional councils are general planning authorities.

2.9 Section 23 of the 1971 Act (Section 20 of the 1972 Act in Scotland) requires planning permission to be obtained for development which is defined in Section 22 (19 Scotland) as "the carrying out of building, engineering, mining and other operations, in, on, over or under land, or the making of any material change in the use of any buildings or other land". Section 22 (19 Scotland) also states "for the avoidance of doubt" that "the deposit of refuse or waste materials on land involves a material change in the use thereof, notwithstanding that the land is comprised in a site already used for that purpose, if either the superficial area of the deposit is thereby extended, or the height of the deposit is thereby extended and exceeds the level of the land adjoining the site". The use of land for the disposal of waste therefore usually requires planning permission. However, Article 3 of the Town and Country Planning General Development Order 1977 (as amended) or Town and Country Planning (General Development) (Scotland) Order 1981, provides that a wide range of development is "permitted" and may be undertaken without a specific grant of planning permission. This applies to the deposit of waste on land in the circumstances set out in Table 2.9.

2.10 It should be noted that the General Development Order (GDO) permission for developments in Table 2.9 may be withdrawn by a direction issued by the Secretary of State or by the planning authority with his approval under Article 4 of the GDO. Express planning permission for a particular development would then be required. The direction cannot, however, apply to development already started when the direction becomes effective. GDO permission can also be withdrawn by means of a condition imposed when planning permission is granted for a development.

Planning applications

2.11 Planning applications relating to the use of land for waste disposal are made to the appropriate authority as indicated in Table 2.4. The applications must be on a form provided by the local planning authority and be accompanied by a plan sufficient to identify the land concerned and by the other plans and drawings and supporting statements necessary to describe the development proposed. In the case of proposals for landfill development it is suggested that the applicant has informal discussions with the planning authority together with the appropriate water authority before submitting an application. Applicants are also strongly advised to consult the Divisional Surveyor of the relevant Agriculture Ministry whenever an application involving agricultural land is to be made. The application should, as far as possible, be supported by the results of site investigations and an assessment of possible impacts on the environment, amenity and where relevant agriculture. It should be noted that a similar need for informal consultations and for site investigations arises in connection with disposal licensing (see paragraph 2.52) and landfill gas control (see Annex 2). Also required in England and Wales, by Section

Table 2.9 Permitted developments as defined by Schedule 1 of the General Development Order

a. The deposit by an industrial undertaker of waste material resulting from an industrial process on any land comprised in a site which was used for such deposit in July 1948 (Class VIII (VII Scotland) para 2).

b. Similar deposits by Local Authorities (Class XIII (XI Scotland) para 2). (There are differences in detail in the provisions of these classes between England and Wales and Scotland and reference should be made to the appropriate General Development Order.)

c. Works carried out by a highway authority for or incidental to the maintenance or improvement of existing highways on land abutting the boundary of the highway (Class XIV (XII Scotland)).

d. The maintenance or improvement of any watercourse or drainage work by a drainage authority, water or hydraulic power undertaking or water authority (Class XVI(b) (XIII Scotland), XVIII (XV Scotland) B2 and C(ii)).

e. The use of any land for the spreading of dredgings by a dock, pier, harbour, water transport, canal or inland navigation undertaking (Class XVIII B (XV Scotland)).

f. The deposit of refuse or waste material on operational land by a number of statutory undertakers (Class XVIII (XV Scotland)).

g. The deposit by mineral undertakers of refuse or waste materials in excavations made by them and already lawfully used for that purpose so long as the height does not exceed the level of adjoining land (Class XIX (XVI Scotland), para 3).

h. The deposit of waste resulting from colliery production activities on land comprising a site used for the deposit of waste materials on 1 July 1948 (Class XX (XVII 3 Scotland)).

References in brackets are to Schedule 1 of the GDO.

A review of the GDO is underway and a number of changes are in prospect.

26 of the 1971 Act and Article 8 of the GDO, is evidence that the application has been advertised both in the local newspaper and by means of site notices. In Scotland, the relevant provisions are Section 23 of the 1972 Act and Article 7 of the GDO as substituted by the General Development (Scotland) Amendment Order 1984. This requires the applicant to notify neighbours of the proposal and the planning authority to advertise it in a local newspaper.

2.12 Article 7(6) (19(5) Scotland) of the GDO provides that a decision on an application for planning permission shall be given within eight weeks (two months Scotland) from the date the valid application is received, unless the period has been extended by agreement between the applicant and the local planning authority. Article 7A of the GDO applies the eight week limit to applications for any consent, agreement or approval required by a condition on a planning permission. AN OUTLINE PLANNING PERMISSION CANNOT BE GRANTED FOR A LANDFILL OPERATION.

2.13 In England and Wales, where the development of a landfill site is to be undertaken by a local authority within its own area, it resolves to seek permission for the carrying out of development and that resolution is treated in the same way as a planning application. If, in due course, the local authority proposes to proceed with the development, it passes a second resolution to carry out that development and on passing that resolution, planning permission is deemed to be granted by the Secretary of State. This permission, however, enures for the benefit only of the local authority. A similar procedure is provided for the deemed planning permission for development of land vested in a local authority and within its own area but which the local authority does not itself propose to undertake. The deemed permission in this case runs with the land. In Scotland, different procedures set out in the Town and Country Planning (Development by Planning Authorities) (Scotland) Regulations 1981 (as amended), apply.

Consultations

2.14 Planning authorities are required by the GDO as amended to carry out consultations on planning applications in specified circumstances. The object of such consultations is to establish whether the basic planning presumption in favour of development should be set aside and, if not, what conditions it may be necessary to attach to the grant of planning permission. The following paragraphs summarise the main types of consultations which may be required on an application for a landfill site. The summary is not exhaustive and reference should be made to the GDO for further details.

2.15 *Trunk road and special roads* In England and Wales, where the proposed development is within 67 metres) of the centre line of such a road or involves access to it, the planning authority must notify the Secretary of State under Article 11 of the GDO. The authority is then precluded from determining the planning application until it receives from the Secretary of State either a direction (under Article 10) restricting the grant of planning permission or notification that he does not propose to give such a direction. (Likely to be amended in 1986.) The authority is also required to consult the Secretary of State if it considers that the proposed development would result in a material increase of traffic entering or leaving a trunk road or using a level crossing over a railway [Article 15(1)(b)]. In Scotland the consultation arrangements for trunk or special roads are prescribed at Article 13(1)(a) and (b) of the GDO.

2.16 *Highways other than trunk roads* If the development involves the formation or alteration of the means of access to a highway which is not a trunk road, the local planning authority is required to consult the local highway authority (Article 15(1)(c), (13(1)(c) Scotland)).

2.17 *Neighbouring authorities* Article 15(1)(a) (13(1)(d) Scotland) provides that where development is likely to affect land in the area of another (in Scotland, adjoining) planning authority, the authority responsible for determining the application shall consult with that authority.

2.18 *Areas of National Coal Board working* If the proposal involves the erection of a building (other than temporary) in an area being worked for coal by the NCB, the planning authority is required to consult the NCB (Article 15(1)(d), (13(1)(e) Scotland)).

2.19 *Water considerations* Planning authorities are required to consult their Regional Water Authority in England and Wales, or the River Purification Board in Scotland on all applications to use land for the deposit of any kind of refuse or waste (Article 15(1)(f), 13(1)(i) Scotland).

2.20 *Sites of special scientific interest* The planning authority is required to consult the Nature Conservancy Council where development of land is in an area notified as a site of special scientific interest (Article 15(1)(g), (13(1)(g) Scotland)). DOE Circular 108/77 (150/77 Welsh Office) discusses the considerations relevant to development in or near these sites and also in or near nature reserves which the NCC are empowered to establish.

2.21 *Agricultural land* In England and Wales, the local planning authority must under Article 15(1)(i) consult the Ministry of Agriculture, Fisheries and Food where development, other than for agricultural purposes, would cause the loss of not less than 4 hectares of agricultural land and is not in accordance with the provisions of the development plan. Consultation is also required where the direct loss is less than 4 hectares but further loss is likely to follow. Appropriate advice on the development of agricultural land is given in DOE Circular 75/76 (110/76 Welsh Office and Scottish Development Department 77/75). SDD Circular 24/81 and the associated Town and Country Planning (Notification of Applications) (Scotland) Direction 1981 describe the arrangements for consulting the Department of Agriculture and Fisheries for Scotland, and for notifying applications to the Secretary of State.

2.22 *Safeguard areas* Some types of establishment, by the nature of the activities carried on there, may constitute a danger to any development taking place on neighbouring land or may themselves require to be protected against such development. The Secretary of State has made a number of safeguarding directions under Article 15(4), (13(2) Scotland) of the GDO, requiring the local planning authority to consult the government department or the authority responsible for the establishment, over any development within an area shown on the safeguarding map. There is a general direction relating to aerodromes — the Town and Country Planning (Aerodromes) Direction 1981 (1982 Scotland) published in DOE Circular 39/81 (SDD 16/82) — and consultation is required on all proposed developments, within a 13Km radius shown on the individual safeguarding maps, which might endanger the safety of aircraft by attracting large numbers of birds. Landfill sites will come within this category of consultation.

2.23 *County – district consultation* In England, while the county council is the planning authority for determining waste disposal applications, it is obliged to consult the appropriate district council. From 1 April 1986 in the present GLC and Metropolitan Counties, the planning authority will be the borough or district council (section 3, Local Government Act 1985).

2.24 *District – parish consultation* In England paragraph 20 of Schedule 16 of the Local Government Act 1972 requires a district planning authority to advise parish or community council of any planning application for development of land in their area which the council have expressed a wish to see. Article 17 of the GDO provides that the council may make representations to the determining local planning authority.

2.25 *Period for consultation* A local planning authority required to consult under Article 15 (13 Scotland) of the GDO must allow not less than 14 days for representations

to be made. But in the interests of efficiency in carrying out consultations generally, including those under Article 15, the National Development Control Forum have issued a code of practice that has been agreed by the Secretary of State, the local authority associations and the bodies which planning authorities regularly consult. This calls for completion of consultation as soon as possible within a 28 day period. There is a separate code of practice in Scotland which is described at Annex A of SDD Circular 24/1981.

2.26 *Other interests* In practice, planning authorities may consult a wider range of interests than those mentioned and are encouraged to do so in advice given in DOE/WO/SDD circulars (see Table 2.26). Such interests might additionally include the Health and Safety Executive, statutory undertakers, transport authorities, local amenity groups and residents' associations.

Principles and conditions

2.27 In dealing with applications for planning permission, the planning authority is required to have regard to the provisions of the development plan, so far as it is

Table 2.26 Non statutory consultations

Subject	Source of request to consult	Those to be consulted
1. Land near nuclear power stations	Letter to affected authorities	Health and Safety Executive.
2. Agricultural development	Circular 75/76	Divisional Surveyor MAFF to be consulted only if the local planning authority finds itself unable to determine the application without his appraisal of agricultural facts.
3. Hazardous materials		Health and Safety Executive and Fire Authorities.
4. Noise problems	Circular 10/73	Public health authorities, highway authorities, aerodrome operators.
5. Conservation areas	Circular 23/77	Local authorities are advised to establish Conservation Area Advisory Committees and to refer to them for advice applications which might affect the character or appearance of a conservation area. They are also encouraged to seek the advice and views of local residents and amenity groups when preparing schemes for preservation or enhancement.
6. Development which would affect the setting of a listed building	Circular 23/77	Authorities are asked to ensure that they bring fully instructed opinion to bear on any proposal which is likely to have an adverse effect on the setting of a listed building.
7. Proposals which have an impact on a neighbourhood	Circular 71/73	Local amenity societies, neighbours etc at the authority's discretion.
8. Sports facilities	Circular 33/70	Local or Regional Sports Councils.
9. Pipelines	Circular 115/76	Regional Water Authorities, Statutory water companies etc and, where appropriate the Historic Buildings and Monuments Commission.
10. Land near explosives factories and magazines (Schedules of sites held by Relevant Case Records, planning branches and local authorities concerned)	Letter to affected authorities	Health and Safety Executive.

material to the application, and to any other material considerations. The authority's waste disposal plan will be a material consideration and any land use policies in that plan will normally be incorporated in the development plan. There is a basic planning presumption in favour of development and DOE Circular 22/80 (40/80 WO and SDD Circular 24/81) states that development should be prevented or restricted only where this serves a clear planning purpose and the economic effects have been taken into account. The circular, however, also notes the Government's commitment to conservation and improvement of the countryside, natural habitats and areas of architectural, natural, historical or scientific interest (see also DOE Circular 14/85). Great importance is attached to the maintenance of green belts and the prevention of inappropriate development within them. DOE Circular 22/80 (SDD 24/81) goes on to say that the Government will not allow more than the essential minimum of agricultural land to be diverted to development, nor land of higher agricultural quality to be taken where land of a lower quality could reasonably be used instead.

2.28 Section 29 of the 1971 Act (Section 26 of the 1972 Act, Scotland) empowers planning authorities to grant planning permission "either unconditionally or subject to such conditions as they think fit". The conditions however must relate to the land comprised in the application and must be reasonable and serve a planning purpose. Additionally Section 30 (27 Scotland) enables conditions to be set to regulate the development or use of other land as long as it is under the control of the applicant and the condition is related to the development being permitted. Guidance on the use of conditions is given in DOE Circular 1/85 which suggests that the following tests should be satisfied before a condition is imposed:
Is the condition:

(a) necessary?

(b) relevant to planning?

(c) relevant to the development to be permitted?

(d) enforceable?

(e) precise?

(f) reasonable in all other respects?

In relation to (b), the circular advises that conditions should not be imposed to control matters that are regulated by other statutes (see also 2.64-68).

2.29 Additional provisions contained in Section 5 of the Town and Country Planning (Minerals) Act 1981 (S5: S22 Scotland) may be applied where planning permission for mineral working is also a permission for landfill waste disposal. These powers enable mineral planning authorities to impose "aftercare" conditions in relation to land which is to be restored to agriculture, forestry or amenity use after mineral working has ceased. This is achieved by the addition of a new Section 30A to the 1971 Act (Section 27A of the 1972 Act, Scotland). (See DOE Circular 1/82, Welsh Office Circular 3/82 SDD Circular 5/82).

2.30 Every planning permission is granted or deemed to be granted subject to the condition that development must be begun within a specified period. The period is normally five years from the date of planning permission. However the planning authority can, if it thinks fit, vary the five year period.

Planning appeals and call-in

2.31 The conditions under which an applicant may appeal to the Secretary of State are set out in Sections 36 and 37 of the 1971 Act as amended by the Local Government Planning and Land Act 1980 and Section 53 of the 1971 Act. (For Scotland, Sections 33 and 34 of the 1972 Act as amended by the Local Government and Planning (Scotland) Act 1982 and Section 51 of the 1971 Act.) An appeal may be made if:

(a) planning permission is refused or is granted subject to conditions not acceptable to the applicant;

(b) any consent, agreement or approval required by a condition imposed on a grant of planning permission is refused or is granted subject to conditions;

(c) any approval required under a development order is refused or is granted subject to conditions;

(d) a determination is given under Section 53 of the 1971 Act (Section 51 of the 1972 Act, Scotland) that development is involved or that an application for permission need to be made; or

(e) the authority fails to determine the application within the prescribed period, or fails to give notice to the applicant of the reference of the application to the Secretary of State. The prescribed period is eight weeks (two months in Scotland) or whatever longer period may be agreed by the applicant and the local planning authority.

2.32 Although the majority of appeals are decided on the basis of written representations and by Inspectors (Reporters in Scotland) appointed by the Secretary of State, appeals concerning landfill sites are occasionally the subject of public local inquiry and decision by the Secretary of State. His decision is final and may be challenged only in the High Court (in Scotland, the Court of Session), on the grounds that it is not within the powers of the Act or that any procedural requirement has not been complied with.

2.33 Section 35 of the 1971 Act (Section 32 of the 1972 Act, Scotland) empowers the Secretary of State to call-in any particular planning application for his own decision.

Regulation 7(1) of the Town and Country Planning General Regulations 1976 (in Scotland, Regulation 6 of the Development by Planning Authorities (Scotland) Regulations 1981) gives him a similar power where a local authority is seeking a deemed planning permission for development. DOE Circular 2/81 (Welsh Office Circular 2/81 and SDD Circular 24/81) reiterates that applications will be considered for call in only if planning issues of more than local importance are involved. Applications called in are normally subject to public local inquiry.

Enforcement

2.34 A planning authority is empowered to take enforcement action where development is carried out without planning permission or without compliance with conditions or limitations imposed by the planning permission. The action must, with the exception of material changes of use of land, be taken within four years of the date of the breach of planning control. Material changes of use of land such as waste disposal are not subject to the four year rule. The enforcement notice must specify the steps to be taken to remedy the breach. The notice cannot, however, come into effect until a period of not less than 28 days has elapsed. During this period the person served with the enforcement notice may appeal to the Secretary of State on the various grounds set out in Section 88 (85 Scotland) of the Act. The notice then has no effect until the appeal is determined. Where an enforcement notice has been served but has not taken effect, the local planning authority may serve a stop notice under Section 90 (87 Scotland) prohibiting any activity alleged to be in breach of planning control. The notice cannot take effect until at least three days and not more than 28 days after service, provided that (in England and Wales) the activity did not commence more than 12 months earlier. The 12 month period does not affect an activity if it is, or is incidental to, building, engineering, mining or any other operations or the deposit of refuse or waste material. A person suffering loss or damage directly attributable to a stop notice may claim compensation in the circumstances prescribed in Section 177 (166 Scotland).

Certificates of established use

2.35 A person having an interest in land may apply to the planning authority for a certificate that a particular use of land has become established and is not, therefore, liable to enforcement action. Section 94 (90 Scotland) provides that a use of land is established if:

(a) it was begun before the beginning of 1964 (1965 Scotland) without planning permission in that behalf and has continued since the end of 1963 (1964 Scotland); or

(b) it was begun before the beginning of 1964 (1965 Scotland) under a planning permission in that behalf granted subject to conditions or limitations, which either have never been complied with or have not been

complied with since the end of 1963 (1964 Scotland); or

(c) it was begun after the end of 1963 (1964 Scotland) as the result of a change of use not requiring planning permission and there has been, since the end of 1963 (1964 Scotland), no change of use requiring planning permission.

Section 95 (91 Scotland) provides the applicant with a right of appeal to the Secretary of State and, in England and Wales, empowers the latter to call in applications for his own determination. A Certificate of Established Use does not constitute a basis for obtaining a disposal licence. While such a certificate constitutes a statutory defence against enforcement action, it does not constitute planning permission. Indeed, the prescribed form for such a certificate includes a statement that it is not a grant of planning permission and does not necessarily entitle the owner or occupier of the land to any consequential statutory rights which may be conferred where planning permission has been granted.

Revocation and modification orders

2.36 A planning authority is empowered, where it appears to them expedient to do so, to revoke or modify a planning permission to develop land. An order served under Section 45 (42 Scotland) does not take effect until confirmed by the Secretary of State and any person on whom the order is served has a right to be heard by a person appointed by the Secretary of State. Where these persons do not object to the order, Section 46 (43 Scotland) provides an amended procedure which dispense with the Secretary of State's confirmation. Section 164 (153 Scotland) prescribes the circumstances in which a claim for compensation may be made to the planning authority.

Discontinuance order

2.37 Section 51 (49 Scotland) empowers a planning authority, where it appears to them to be expedient, to require by order that any use of land should be discontinued, or that any conditions should be imposed on the continued use of land, or that any buildings or works should be altered or removed. Once again, the order is subject to confirmation by the Secretary of State and persons on whom the order is served have a right to be heard. Section 170 (159 Scotland) prescribes the circumstances in which a claim for compensation may be made to the local planning authority.

Completion notice

2.38 Where development has begun within the period imposed in the planning permission but has not been completed within the period, the planning authority may serve a completion notice under Section 44 (41 Scotland) if, in their opinion, the development will not be completed within a reasonable period. Those on whom the notice is served have a right to be heard by a person appointed by the Secretary of State. The Secretary of State is required to confirm the completion notice before it becomes effective. At the end of the period specified in the notice (not less than 12 months), the planning permission becomes invalid, insofar as it relates to development not carried out by that time.

Agreements and covenants

2.39 Legally binding agreements between local planning authorities and developers to secure matters which cannot be achieved by conditions attached to a planning permission are becoming more widely used and may have some relevance to the question of providing for the aftercare of landfill sites (see paragraph 2.67 and 6.103). Such agreements by their very nature must be freely entered into. A planning authority cannot make planning permission conditional upon an agreement being entered into (see Appendix A, paragraph 3, DOE Circular 22/83).

2.40 The power most widely used by a local authority is Section 52 of the 1971 Act (Section 50 of the 1972 Act, Scotland). This provides that a planning authority may enter into an agreement with any person interested in land in their area for the purpose of restricting or regulating the development or use of that land. It also provides that the agreement may be enforced by the planning authority against persons deriving title from those with whom the agreement was made as if the authority were possessed of adjacent land and as if the agreement had been expressed to be made for the benefit of that land. The planning authority would therefore be in the same position, in relation to enforcement of the agreement as a person entitled to the benefit of a restricted covenant. The agreement would of course be enforceable against the original covenanter.

2.41 Section 111 of the Local Government Act 1972 (Section 69 of the 1973 Act, Scotland) provides that a local authority may do anything — and this would include making an agreement which is calculated to facilitate or is conducive or incidental to the discharge of any of their functions. Thus the purpose of an agreement under this section could be wider than the purpose of agreements under Section 52 of the 1971 Act (Section 50 of the 1972 Act, Scotland), which can only restrict or regulate the development or use of land.

2.42 Agreements under Section 52 and Section 111 suffer the disadvantage that any positive convenants contained therein are not enforceable against successors in title. This disadvantage can be removed by invoking Section 33 of the Local Government (Miscellaneous Provisions) Act 1982 which replaced Section 126 of the Housing Act 1974. Section 33 secures for agreements made by a local authority, under whatever power, the enforceability of positive covenants in the agreement against successors in title. It also gives the local authority default powers to enter the land and do what the covenant requires.

2.43 There are also powers in a number of local Acts enabling local authorities to secure agreements but most of these will cease to be effective at the end of 1986 by virtue of Section 262(9) of the Local Government Act 1972.

Pollution control legislation

Control of Pollution Act 1974

2.44 The Control of Pollution Act 1974 is the principal legislation regulating and controlling the disposal of waste. It affects landfill operations in three main ways. First, it requires the disposal authority to prepare a waste disposal plan for its area. These plans, among other things, will examine the role that landfill might play in the authority's waste disposal strategy. Second, the Act requires a landfill site to be licensed before operations can commence. This licence, which can be revoked or modified, sets the conditions within which the landfill operation will be conducted. Third, the Act makes provision for the control of discharges to waters. While these matters are dealt with during disposal licensing consideration, provision is made in the Act for consents to be required from the relevant water authority.

Waste disposal plans

2.45 Section 2 of the 1974 Act requires each waste disposal authority to produce a plan for the disposal of waste arising within or entering into its area. It was the intention of Parliament that through the preparation of such plans waste disposal authorities would be well informed about all wastes in their area and be in a position to satisfy themselves that these wastes could be handled and disposed of safely.

2.46 The waste disposal plan, which should look forward 10 years, aims to produce a cost effective strategy for waste disposal having due regard to the safeguarding of the environment and the use of waste as a resource. The plan should provide a comprehensive appraisal of waste disposal problems and the advantages and disadvantages of alternative disposal methods before setting out the preferred strategy. It should also cover the environmental criteria by which landfill proposals will be judged.

2.47 A disposal authority may choose to show sites diagrammatically in its waste disposal plan and identify areas where sites could be located, rather than specific sites. This course may suit an authority that has still to acquire a particular site to meet its own needs, but the possibility of thereby blighting large tracts of land must be considered.

2.48 Some county planning authorities are finding it useful to prepare waste disposal subject plans, which are local plans prepared under the Town and Country Planning Act 1971 as amended (1972 Scotland) (see paragraphs 2.4 and 2.5). These set out land use policies and proposals for the disposal of waste. They complement waste disposal plans prepared under the Control of Pollution Act 1974. Consultation between waste disposal and planning officers is desirable in deciding on the best approach for a particular area.

Disposal licensing

2.49 With only limited exceptions (see below) Part 1 of the 1974 Act, required that all waste disposal operations (including landfill sites) taking controlled wastes are licensed or covered by resolution by the disposal authority (see Table 2.4). Section 30 of the Act describes 'controlled waste or any such waste' as covering household, industrial and commercial wastes, these being defined as follows:

(a) household waste consists of waste from a private dwelling or residential home or from premises forming part of a university or school or other educational establishment or forming part of a hospital or nursing home;

(b) industrial waste consists of waste from any factory within the meaning of the Factories Act 1961 and any premises occupied by a body corporate established by or under any enactment for the purpose of carrying on under national ownership any industry or part of any industry or any undertaking, excluding waste from any mine or quarry; and

(c) commercial waste consists of waste from premises used wholly or mainly for the purpose of a trade or business or the purpose of sport, recreation or entertainment excluding:

(i) household and industrial waste and

(ii) waste from any mine or quarry and waste from premises used for agriculture within the meaning of the Agriculture Act 1947 or in Scotland the Agriculture (Scotland) Act 1948 and

(iii) waste of any other description prescribed for the purpose of this sub-paragraph.

The disposal of certain controlled waste has been exempted from licensing requirements. These exemptions are described in Table 2.49.

2.50 The basic aim of licensing is to ensure that landfilling operations entail no acceptable risk to the environment and to public health, safety and amenity. Given that planning permission has been received for a proposed landfill development (or that the development is

Table 2.49: Controlled wastes exempt from licensing*

a. building demolition etc waste deposited or disposed of on a demolition site, site being or about to be used for construction, improvement or repair etc of a building with the occupier's consent

b. spent railway ballast deposited on operational land belonging to the British Railways Board

c. dredging waste which is deposited on the banks of a watercourse

d. waste from maintaining any park, sportsfield, public garden or recreation ground which with the occupiers consent is deposited or disposed within the grounds where it originated

e. waste deposited as part of research into its effect on the environment or research into plant etc designed to deal with the waste

f. waste deposited directly on land in certain circumstances and with the occupiers consent for a period of not more than 1 month

g. waste deposited in a designated receptacle the contents of which is subsequently to be disposed of elsewhere

h. waste disposed of on site on which it is produced by a static plant including incinerators with a disposal capacity of not more than 200 kilogrammes per hour

j. waste disposed of as an integral part of the industrial process that produces it

Apart from item j, none of the above cases is exempted if the waste is poisonous, noxious or polluting and may give rise to an environmental hazard if deposited on the land.

* These exemptions applied at the date of publication of this report but may be changed under the proposals for amendment of the Control of Pollution Act (Licensing of Waste Disposal) Regulations 1976 and 1977 (see paragraph 2.2).

exempted: see paragraph 2.9) an application for a license can be rejected only on the grounds of danger to public health or risk of water pollution. Consequently, licences will almost certainly be issued with conditions attached, these conditions being intended to ensure that health and environmental quality, and in particular water quality, are safeguarded. Such conditions are intended to regulate the operation of the landfill in some detail (see paragraph 2.57) and the drawing up of a disposal licence is therefore primarily a technical task in which many factors have to be considered. Disposal licences thus provide landfill operators with clear directions on the operating standards they are expected to achieve by the responsible control authorities. Disposal licensing also has other, more general, functions. It provides a mechanism by which the responsibility for deciding the conditions to be imposed at a landfill site is placed at a suitable local level, so that proper account can be taken of local circumstances. Also in conjunction with waste disposal plans (see paragraph 2.45 et seq) and other relevant policy documents, it can provide disposal authorities with information to assist them in discharging their general responsibilities to ensure that adequate waste disposal facilities continue to be available; this assumes that licence conditions require that records are kept and provided to the waste disposal authority on all waste disposal activities and not just those wastes subject to Section 17 of the Control of Pollution Act 1974. Licensing also enables the controlling authorities to ensure that changing patterns of landfill practice do not prejudice safe operations and, equally, that landfill operators take full advantage of technical progress.

Obtaining a disposal licence

2.51 The procedure to be followed in order to obtain a disposal licence can be divided into four stages:

(a) Informal discussions with authorities,

(b) Submitting a licence application,

(c) Statutory consultations,

(d) Granting a licence.

An outline guide to obtaining a disposal licence is given in Figure 2.51.

2.52 *Informal discussions* It is advisable for a prospective site operator to hold informal discussions with relevant parties before submitting a licence application (or planning application: see paragraph 2.11). The parties involved could include the disposal authority, the Regional Water Authority in England and Wales, or River Purification Board in Scotland in whose area the site is located and others such as the local district council (in England) and the divisional office of the relevant Ministry or Department of Agriculture, Fisheries and Food. Such discussions should normally be based on information obtained from site investigations, although at this early stage the results of an initial reconnaissance study may be sufficient (see paragraph 3.106).

2.53 *Submitting a licence application* If, in the light of informal discussion and the results of site investigations, the applicant decides to proceed, the next step is submission of a formal application for a disposal licence. In support of this application the disposal authority will require details of any planning permission or exemption. The disposal authority will also require the applicant to supply the results of a properly designed and conducted site investigation together with information about the way in which the site is to be developed and operated — the working plan. A detailed description of the matters that may need to be covered in the site investigation and in the working plan is given in paragraphs 3.119 and 3.120 respectively. The extent and rigour of the site investigation will depend on the characteristics of the particular site and the nature of the wastes that it is proposed to deposit at the site; in most cases professionally executed site investigations and landfill design will be required. It should be noted that the results of site investigations may be available at the time of the planning applications given

that, increasingly, planning and disposal licence applications are made at the same time. In such cases the results should be used to support both applications.

2.54 *Statutory consultations* Section 5(3) of the Control of Pollution Act 1974 requires the waste disposal authority to grant a disposal licence unless:

(a) a valid planning permission or its equivalent has not been granted,

(b) rejection is necessary to prevent danger to public health, or

(c) rejection is necessary to prevent pollution to water.

It should be noted that although damage to amenity does not provide grounds for rejection of an application for a disposal licence, a licence could subsequently be revoked on the grounds that such damage could not be avoided (see paragraph 2.59).

2.55 In determining whether a proposed operation is likely to endanger health or cause water pollution the disposal authority must consult the Regional Water Authority (River Purification Board in Scotland) and, in England, with the waste collection authority in whose area the site is located. The disposal authority is also advised to consult the Health and Safety Executive; it should also consult the British Geological Survey in cases involving waste disposal in shafts, galleries, wells, or boreholes, or liquid disposal into fractures, fissures or intergranular pore-spaces in geological formations by borehole. The disposal authority may also seek the views of other parties if it so wishes but there is no requirement to advertise a disposal licence application or its transfer (cf paragraph 2.11).

2.56 Provided there are no objections on the grounds of potential water pollution or danger to public health, a licence must be issued, albeit with appropriate conditions attached. However, if objections are raised on public health grounds then it is up to the disposal authority to decide if a licence should be issued. If a Regional Water Authority (River Purification Board in Scotland) requests the disposal authority not to issue a licence or disagrees with the disposal authority as to the conditions to be specified in the licence either of them may refer the matter to the appropriate Secretary of State, and the licence will not be issued except in accordance with his decision. Once a licence is issued a consent from the Water Authority (River Purification Board in Scotland) must be obtained before any leachate or other liquid from a landfill site is discharged to sewer, watercourse or tidal water.

Conditions on disposal licence

2.57 Conditions contained in a disposal licence will reflect local circumstances as well as common landfill

Figure 2.51 AN OUTLINE GUIDE TO OBTAINING A DISPOSAL LICENCE

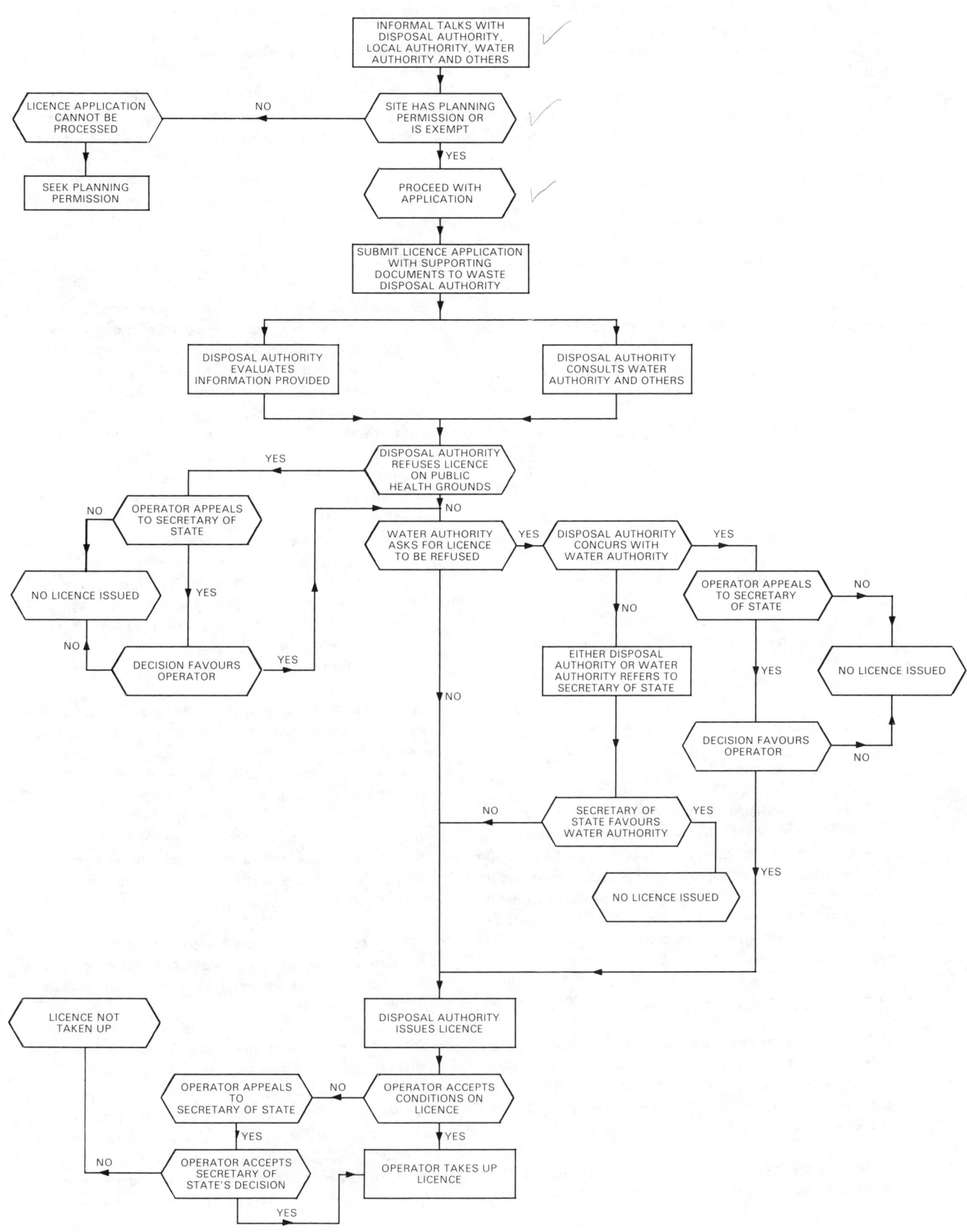

standards which the disposal authority expect to be achieved. Such conditions are determined after detailed technical consideration of the wide range of facets that make up a modern landfill operation. Each landfill site calls for individual treatment and differing conditions are likely to be set for each licence. However, the considerations may include:

(a) timing of operations including the phasing of the development not covered by planning conditions,

(b) type of material and rate of fill,

(c) working hours for landfilling and restoration,

(d) phasing of operations including limits on areas to be landfilled,

(e) progressive restoration as set out in the working plan,

(f) direction of landfilling and extent of tipping face,

(g) stripping and storage of top soil and sub-soil,

(h) measures to deal with water pollution and leachate,

(i) fencing and gates,

(j) layering of waste and use of compaction equipment,

(k) cover materials both intermediate and final,

(l) dealing with gas emissions and/or migration,

(m) monitoring and,

(n) keeping records.

The list is not exhaustive; it is intended simply to convey an idea of the wide area that licence conditions may cover. These conditions do, however, apply only during the life of the site and do not give control over post-closure management (or as regrettably occurs occasionally, abandonment). This matter is being considered, together with the relevance of planning controls in the post-closure phase of operations.

Disposal licence appeals

2.58 An applicant may appeal to the Secretary of State if he feels that the disposal authority has acted unfairly or unreasonably. Four circumstances are provided for by the Control of Pollution Act 1974 in which an appeal may be made. These are:

(a) an application for a licence or modification of an existing licence is rejected or deemed to be rejected,

(b) objections are raised to a condition(s) contained in a licence,

(c) conditions in an existing licence are modified, or

(d) a disposal licence is revoked.

There is also the special case mentioned above (paragraph 2.56) where, in the event of a disagreement between a disposal authority and a Regional Water Authority (River Purification Board) over issuing a disposal licence or over the conditions to be placed on a licence, either party may refer the matter to the Secretary of State for determination.

Enforcement

2.59 A disposal authority is required to supervise all sites it licences to ensure that licence conditions are being observed and continue to be appropriate. Such supervision should ensure that landfill operations do not result in water pollution, endanger public health or cause a serious detriment to local amenities. Supervision is carried out by visits to the site by disposal authority Inspectors. Where a breach of licence conditions is observed or when the operation is considered to pose a threat to water, health or amenity, a number of enforcement powers and duties are available to the disposal authority under the 1974 Act. In summary these are:

(a) to prosecute under Section 3 anyone operating a landfill site without a valid licence, depositing waste contrary to licence conditions, or using plant or equipment for the purpose of disposing of waste contrary to licence conditions,

(b) to revoke under Section 7(4) any licence where continuation of activities would be likely to cause water pollution, or serious detriment to local amenities and where such pollution or detriment cannot be avoided by changing the licence conditions,

(c) to serve notice under Section 9(4) for the licence holder to comply with a condition within a given time and failing this give further notice of intended licence revocation,

(d) to serve notice under Section 16(1) that, in order to eliminate or reduce the consequences of a deposit of waste, such waste must be removed or specified steps are to be taken,

(e) to modify the licence under Section 7(1) so as to meet changed conditions.

2.60 A disposal authority and a collection authority also has powers under Section 16 of the Act to remove from any unoccupied land wastes deposited illegally so as to remove or prevent water pollution or danger to public health and under Sections 16(6) and 88 to recover the cost from the person responsible. In most cases the 1974 Act also specifies defenses against actions taken by a disposal authority.

Sites operated by a local authority

2.61 If a disposal authority itself proposes to undertake landfill operations the relevant council, of which it is a part, must pass a resolution including a specification of the conditions in accordance with which the landfill operation is to be conducted.

2.62 The disposal authority is required to submit details of its proposed operation and disposal licence to the local Water Authority (River Purification Board) and also the collection authority in whose area the proposed site is located. In Scotland where the disposal authority is not a district planning authority, the details should also be submitted to the general planning authority. If the relevant water authority requests the disposal authority not to proceed with the proposal or disagrees with the conditions to be specified in a resolution, the proposal may not proceed and, as with disposal licences for private sector sites, can be referred to the appropriate Secretary of State for determination.

2.63 A resolution passed by a local authority may also be varied and rescinded by a subsequent resolution. It is the duty of the disposal authority to regulate its own operations as it would those of the private sector and thereby maintain consistent standards. The Hazardous Waste Inspectorate can advise local authorities on these matters.

Relationship between planning and disposal licensing

2.64 Planning control and disposal licensing are intended to work side by side. In making planning permission a prerequisite to issuing a disposal licence, the Control of Pollution Act 1974 assumes that planning matters relevant to the suitability of a site for waste disposal will have been considered and appropriate conditions made. Thus the primary consideration in issuing a disposal licence should be to control operations in order to prevent water pollution and danger to public health. Current advice on the relationship between planning and waste disposal legislation is given in DOE Circular 55/76 (Welsh Office 76/76) and Waste Management Paper No 4 which is to be revised.

2.65 The 1974 Act does not detract from any of the powers given under the Town and Country Planning Act 1971 (Town and Country Planning (Scotland) Act 1972) and is particularly relevant to wastes excluded from the scope of the Control of Pollution Act 1974. In considering the siting of a proposed development, therefore, prevention of pollution of water and dangers to public health can equally be a matter for development control under the 1971 (1972) Act as it is for licensing control under the 1974 Act. For example, in many cases consideration of the different types of waste acceptable at a

particular site will be dealt with by way of a condition in a waste disposal licence. However, if the local planning authority consider that such a condition is material to the granting of planning permission — if, for example they are prepared to grant permission only for specified types of waste — then it is open to them to impose such a condition.

2.66 Nevertheless it remains a matter of sensible administration that planning control is not normally used to secure objectives for which provision is made in other specific legislation. Where matters of close concern to planning are also regulated by other statutes or by common law it is usually an undesirable duplication to seek to control such matters by attaching a condition to a planning permission. Consequently unless it forms a wider, material planning consideration, the detailed operation of a landfill should be regulated by disposal licensing.

2.67 One of the main distinguishing features between planning legislation and disposal licensing provisions is that a planning permission and its conditions enures for the benefit of the land (except where the development is by a planning authority: see paragraph 2.13), whereas disposal licence conditions apply only for the duration of a licence and are more readily varied. The enforcement powers of the 1974 Act apply mainly to the licence holder, not to the land-owner. Continued access to boreholes for monitoring purposes or maintenance of leachate control systems, which may need attention for several decades after completion, cannot therefore be provided for in disposal licensing conditions (but see paragraph 2.2). The position is similar for restoration or aftercare once waste disposal operations have ceased and the licence has been cancelled. Some planning authorities have attached conditions to recent planning permissions which aim to ensure that restoration and after care controls will apply after expiry of a disposal licence. There are some difficulties in this approach, particularly in regard to enforcement of conditions for continued monitoring and control of methane and leachate escaping from landfill sites after restoration. Some authorities are adopting alternative approaches such as asking applicants to include post restoration pollution prevention measures in their planning applications, or by making Section 52 Agreements under the 1971 Act (see paragraphs 2.39-2.42 and 6.103).

Special waste

2.68 Some wastes are also subject to the specific controls set out in the Control of Pollution (Special Waste) Regulations 1980 (SI 1980: 1709). These regulations are based on the premise that some wastes are sufficiently hazardous to present a threat to human health if encountered during their carriage between the point of arising and the waste disposal facility. A record of their final place of deposit is also to be kept. Except for this record keeping aspect, these regulations are not directed at the operation of the disposal facility. Rather they specify an information system where, by means of a prescribed consignment note requiring signature at each transfer of the waste, the waste can be followed from the producer to the final disposer. Additionally the regulations require the disposer's disposal authority to be notified at least three clear days (and not more than one month) before removal of the waste from the producer unless the waste has been granted regular consignment status by the producer's disposal authority. On receipt of the waste the disposer must complete a copy of the consignment note and send it to the producer's disposal authority.

2.69 A controlled waste (see paragraph 2.49) is defined as 'Special' if:

(a) it consists of or contains any of the substances listed in Part I of Schedule 1 of the Regulations (and reproduced in Table 2.69) and by reason of the presence of such substances

(i) is dangerous to life within the meaning of Part II of Schedule 1, or

(ii) has a flash point of 21 degrees Celsius or less as determined by the methods and apparatus laid down by BSI in BS 3900: Part AV8 1976 (EN53), or

(b) it is a medicinal product, as defined in Section 130 of the Medicines Act 1968, and is available only in accordance with a prescription given by an appropriate practitioner as defined in Section 58(1) of that Act.

2.70 A landfill operator accepting 'Special Wastes', unless he is also the disposal authority, is required to keep at the site a register containing copies of all consignment notes. The register must be maintained until the disposal licence is surrendered or revoked when it must be handed over to the disposal authority. The landfill operator is also required to record the location of each deposit of 'Special' waste and again keep these records until the disposal licence is surrendered or revoked and then similarly send them to the disposal authority. Such records can comprise either a site plan marked with a grid, or a site plan with translucent overlays on which deposits are shown in relation to the contours of the site. The deposits should be described in the location records by reference to the register of consignment notes, except where the waste is disposed of by pipeline, or within the curtilage of a factory or other premises at which it is produced, or is disposed of by the disposal authority at a site in its own area. In these circumstances, the deposits should be described by reference to a record of the composition of the waste and the date of disposal.

The Public Health Act 1936

2.71 The disposal of waste to land, unless carried out expeditiously, can cause nuisance which could be covered

Table 2.69 'Special' wastes — listed substances

Control of Pollution (Special Waste) Regulations 1980
SI 1980 No 1709 — Schedule 1, Part I

Acid and alkalis

Antimony and antimony compounds

Arsenic compounds

Asbestos (all chemical forms)

Barium compounds

Beryllium and beryllium compounds

Biocides and phytopharmaceutical substances

Boron compounds

Cadmium and cadmium compounds

Copper compounds

Heterocyclic organic compounds containing oxygen, nitrogen or sulphur

Hexavalent chromium compounds

Hydrocarbons and their oxygen, nitrogen and sulphur compounds

Inorganic cyanides

Inorganic halogen-containing compounds

Inorganic sulphur-containing compounds

Laboratory chemicals

Lead compounds

Mercury compounds

Nickel and nickel compounds

Organic halogen compounds, excluding inert polymeric materials

Peroxides, chlorates, perchlorates and azides

Pharmaceutical and veterinary compounds

Phosphorous and its compounds

Selenium and selenium compounds

Silver compounds

Tarry material from refining and tar residues from distilling

Tellurium and tellurium compounds

Thallium and thallium compounds

Vanadium compounds

Zinc compounds

A technical memorandum providing guidance on the interpretation of the definition in both general terms and as it relates to the listed substances is given in Waste Management Paper No 23, "Special Wastes: A technical memorandum providing guidance on their definition" (published by the Department of the Environment). 1981. The list is kept under constant review and changes can be expected.

by "Statutory Nuisance" legislation. Section 92 1(c) of the Public Health Act 1936 may describe the nuisance experienced in some cases:

"any accumulation or deposit which is prejudicial to health or a nuisance"

Section 92 1(d) may also apply:

"any dust or effluvia or stream caused by any trade, business, manufacture or process and being injurious or likely to be injurious to public health or a nuisance".

If the nuisance experienced is covered by either of these definitions legal proceedings may ensue in the following way:

A resident who is suffering from the nuisance will initially contact the Environmental Health Officer in the Local Authority. If the Council is satisfied that conditions prevailing amount to a statutory nuisance, it must serve a notice on the person responsible requiring abatement of the nuisance and specifying steps to be taken to that end. Such a notice is enforceable. A notice may also be served prohibiting recurrence of the nuisance. Section 99 of the 1936 Act allows an individual to take an alleged offender to court directly, if the Council is unwilling to do so.

When proceedings are brought, a company may defend its action by proving that "best practicable means" have been taken to prevent the nuisance, or counteract its effect.

If a local authority is not satisfied that a nuisance has been abated, it may take proceedings against a company in the High Court under Section 100 of the Public Health Act 1936 to secure the abatement of the nuisance. In practice this could mean that the company may be faced with closure. The best practicable means defence is not valid when action is taken under Section 100.

Health and Safety

Health and Safety at Work etc Act 1974

Requirement and duties of employers

2.72 The requirements of the Health and Safety at Work etc Act 1974 apply to most employment activities. In addition to actual landfill operations, they cover such ancillary activities as site investigation, site preparation and site restoration. The Act places a general duty on employers to ensure, so far as is reasonably practicable, the health, safety and welfare at work of all their employees. This duty is extended to include, so far as is reasonably practicable:

(a) the provision and maintenance of plant and

systems of work which are safe and without risk to health;

(b) arrangements for ensuring safety and absence of risks to health in connection with the use, handling, storage and transport of articles and substances;

(c) the provision of such information, instruction, training and supervision as is necessary to ensure the health and safety at work of their employees;

(d) the maintenance of any place of work under the employer's control in a condition that is safe and without risks to health and the provision and maintenance of access to and from the place of work which is safe and without risk to health; and

(e) the provision and maintenance of a working environment for their employees that is safe, without risks to health, and adequate as regards facilities and arrangements for their welfare at work.

Moreover, the Act requires an employer who employs five or more persons to prepare (and to revise when appropriate) a written statement of his general policy with respect to the health and safety at work of his employees and the organisation and arrangements which exist to implement the policy. The employer is required to bring the statement (and any revision of it) to the notice of all his employees.

2.73 In addition to the duties of employers to make provision for the health, safety and welfare of their employees, the Health and Safety at Work etc Act also requires an employer (and the self-employed) to conduct his undertakings in such a way as to ensure, so far as is reasonably practicable, that persons not in his employment who may be affected by his activities are not thereby exposed to risks to their health or safety. The Act also requires persons having control of non-domestic premises which are made available for use as a place of work by others who are not their direct employees, to ensure, so far as is reasonably practicable, that the means of access to and from the premises and any plant or substance on the premises (or provided for use there) are safe and without risk to health. In the case of landfill operations, these requirements might be regarded as placing duties on site operators in relation to safeguarding such people as members of the public in the vicinity of the site and others who are not the direct employees of the operator but who are required to have access to the site, from dangers arising from activities at the site.

Requirements and duties of employees

2.74 The Health and Safety at Work etc Act 1974 also places a duty on the individual employee to take reasonable care for his own health and safety and that of others who may be affected by what he does, or fails to do, at work. The Act also requires employees to co-operate with

17

their employers so far as is necessary to enable employers to perform the duties and requirements placed on them by Health and Safety legislation.

Other regulations and acts

2.75 Landfill site operators should be aware that the following regulations have been made under the Health and Safety at Work etc Act 1974 and will apply to landfill sites. Other regulations may apply also; advice may be obtained from Health and Safety Executive local offices.

(a) The Safety Signs Regulations 1980 (SI 1980 No 1471), which require that any safety sign giving specific health or safety information (other than signs used for regulating traffic, which should comply with the Road Traffic Regulation Act 1967) should comply with BS 5378, "Safety Signs and Colours".

(b) The Health and Safety (First Aid) Regulations 1981 (SI 1981 No 917), which require the provision of adequate first aid facilities and equipment.

(c) The Control of Lead at Work Regulations 1980 (SI 1980 No 1248), which require steps to be taken to ensure adequate control of occupational exposure to lead.

(d) The Notification of Accidents and Dangerous Occurrences Regulations 1980 (SI 1980 No 804), which require the reporting to the enforcing authority of a fatal injury or "major injury" to any person, or of a dangerous occurrence, if these arise out of or in connection with work. The Regulations give a definition of the term "major injury" and contain, in Schedule 1, a list of the dangerous occurrences which must be reported. The Regulations also require a record to be kept of any fatal or major injury accident, any dangerous occurrence, or any accident at work to an employee which results in more than three days absence from work.

2.76 Proposals to introduce regulations to deal with the control of occupational exposure to substances hazardous to health, the risks arising from electrical installations, and several other important aspects of health and safety have been published by the Health and Safety Commission. Details can be obtained from Area Offices of the Health and Safety Executive or from the Executive at St Hugh's House, Stanley Precinct, Bootle, Merseyside, L20 3QY.

2.77 In addition to the Health and Safety at Work etc Act and the regulations made under it, some landfill sites may be subject to the requirements of the Factories Act 1961 or the Offices, Shops and Railway Premises Act 1963. For example, plant and vehicle repairs will attract the application of the Factories Act and its associated regulations, which deal with matters as diverse as the safe use of abrasive wheels and the storage of highly flammable liquids. Offices in which people are engaged in office work for more than 21 hours per week will be subject to the Offices, Shops and Railway Premises Act 1963.

Enforcement

2.78 The provisions of the Act and associated legislation are usually enforced at landfill sites by inspectors of the Health and Safety Executive. Inspectors' enforcement powers include the issue of letters of caution; the serving of Improvement Notices (requiring contraventions of the law to be remedied within a specified period of time) and Prohibition Notices (requiring the cessation of activities which involve a risk of serious personal injury); and prosecution. Enforcement is not, however, the sole function of the Health and Safety Executive and inspectors are ready to give advice and information to employers on health and safety matters.

Other legislation

2.79 While the Acts and regulations so far discussed in detail provide the main legislative control over waste disposal by landfill, there are a number of other statutory enactments which may impose general controls on landfilling or its associated activities, or more specific controls dependent upon the nature, scale and location of the activity. These include:

Local Government Act 1985

Food and Environment Protection Act 1985

Energy Act 1983

Litter Act 1983

Local Government (Miscellaneous Provisions) Act 1982

Oil and Gas Enterprise Act 1982

Town and Country Planning (Minerals) Act 1981

Wildlife and Countryside Act 1981 (1985)

Local Government Planning and Land Act 1980

Highways Act 1980

Water (Scotland) Act 1980

Refuse Disposal (Amenity) Act 1978

Reservoirs Act 1975

Road Traffic Act 1974

Housing Act 1974

Dumping at Sea Act 1974

Heavy Commercial Vehicles (Controls and Regulations) Act 1973

Supply of Goods (Implied Terms) Act 1973

Water Act 1973

Weeds Act 1973

Gas Act 1972

Gas Safety Regulations 1972

Defective Premises Act 1972

Road Traffic Act 1972

Dangerous Litter Act 1971

Oil in Navigable Waters Act 1971

Public Health (Recurring Nuisances) Act 1969

Mines and Quarries (Tips) Act 1969

Clean Air Act 1968

Countryside Act 1968 (1981 and 1985)

Trade Descriptions Act 1968

Farm and Gardens Chemicals Act 1967

Road Traffic Regulations Act 1967

Rivers (Prevention of Pollution (Scotland) Act 1965

Airport Authorities Act 1965

Scrap Metal Dealers Act 1964

Water Resources Act 1963

London Government Act 1963

Pipelines Act 1962

Public Health Act 1961

Rivers (Prevention of Pollution) Act 1961

Factories Act 1961

Noise Abatement Act 1960

Clean Rivers (Estuaries and Tidal Waters) Act 1960

Radioactive Substances Act 1960

Occupiers Liability Act 1957

Clean Air Act 1956 (1968)

Mines and Quarries Act 1954

Agriculture (Poisonous Substances) Act 1952

Public Health (Drainage of Trade Premises) Act 1937

Petroleum (Consolidation) Act 1928

Alkali etc Works Regulation Act 1906

In addition there are a number of Local Acts whose provisions may impinge on landfills in the areas to which they apply.

CHAPTER 3: SITE ASSESSMENT AND DESIGN

Introduction

3.1 Site assessment is an essential part of the initial stages of landfill development. At this stage also, the prospective operator will need to satisfy himself on the financial and technical viability of the proposed operation, having regard to the characteristics of the projected site, the nature and quantities of the wastes he expects to receive, and the requirements relating to environmental protection which are likely to be imposed by the controlling authorities. Subsequently, if the project proceeds, site assessment can be expanded into a detailed site investigation, leading to the formulation of plans for the design and operation of the landfill which forms an essential part of the disposal licensing procedure. Such an investigation includes technical and scientific assessment of hydrological and hydrogeological features and soil conditions, of physical, chemical and biological factors relating to the ability of the site to handle the wastes, and of the potential impact of the operation on the environment and local community.

3.2 The principal aim of disposal licensing, and hence of site assessment, both in the short and long-term, is to safeguard human health, water quality and amenity from adverse effects due to landfill development. In this chapter the various considerations involved in achieving this aim are first presented, under the headings water quality, landfill gas and general environmental effects. There then follows a discussion of the site assessment process, with particular reference to the landfill developments. An account is then given of the scope of the site investigation and the procedures involved. Finally the content of the working plan is described; this plan is the central document for planning and disposal licence applications and also the blueprint for eventual operation of the site.

3.3 It should be noted that this chapter is concerned with the various factors that apply generally to the assessment and design of landfill sites. Certain special considerations that arise for landfills taking difficult wastes are discussed in Chapter 7.

Landfill and water quality

Introduction

3.4 The infiltration of rainfall and surface water into a landfill coupled with biochemical and physical breakdown of wastes produces a liquor or "leachate" high in suspended solids and with a high organic and inorganic content. If this leachate enters surface or groundwaters before sufficient attenuation has occurred (see paragraph 3.27 et seq), it will cause pollution. Under Part II of the Control of Pollution Act 1974, it is an offence to cause or knowingly permit poisonous, noxious or polluting matter to enter into surface or underground waters.

3.5 Water authorities as well as waste disposal authorities, have a duty to ensure that landfill operations do not adversely affect water quality. While water authorities will seek to provide maximum protection to waters used for potable supplies they also have to safeguard other water interests. Water in undeveloped aquifers is protected as if they were developed; this is now a requirement (except for permanently unusable waters) following the implementation of the 1980 EC Groundwater Directive by provisions under the Control of Pollution Act. The volume and planned future use of an aquifer are no longer factors for consideration in reaching a decision about issuing a disposal licence. The directive aims to ensure protection of groundwater against pollution by certain substances (see Appendix 7b). To meet their obligations, water authorities have developed aquifer protection practices or policies.

3.6 It should be noted that water authorities in England and Wales (the Department of Agriculture and Fisheries in Scotland) will be concerned with other aspects of landfill developments if these bear on their land drainage responsibilities. This would apply, for example, to development of a landfill on a flood plain or any obstruction of the proper flow in a river. In these cases the authorities may have introduced byelaws restricting landfill operations additional to those under Control of Pollution Act disposal licenses.

Landfill site characteristics

3.7 Few sites are naturally ideal for landfill but those that are not can, with careful planning, be engineered to be suitable, though at a cost. Landfill sites fall into a spectrum in terms of the extent to which leachate is contained within the landfill but it is convenient to distinguish two broad categories, that is, 'attenuate and disperse' sites and 'containment sites' which correspond to two different approaches to the protection of water resources.

Attenuate and disperse sites

3.8 Such sites allow the slow release of leachate from the landfill and rely on various attenuation mechanisms (see paragraphs 3.27-3.32 and Chapter 7) operating within the body of the waste and in the unsaturated and saturated zones of the underlying strata to ameliorate the polluting characteristics of the leachate. The attenuation mechanisms include those of dilution and dispersion which help to decrease the effect of leachate on water resources. Suitable geological formations best suited to such types of sites are those with significantly high amounts of clay minerals present and where leachate movement will be through pores or micro-fissures. Attenuation may be improved at some sites by the use of permeable liners (see paragraph 4.64).

3.9 Some landfill sites allow leachate migration at rates too fast for much attenuation to take place through chemical or biological processes, though dilution and dispersion in the underlying saturated zone will still be important. These sites in their natural state are usually suitable only for relatively inert wastes, but they can be engineered to accept other wastes.

Containment sites

3.10 These sites are intended to isolate wastes and leachate from the environment for a considerable time (perhaps for decades, or even hundreds of years). This aim may be achieved by taking advantage of strata with a permeability low enough to prevent significant seepage of leachate through the base and sides of the landfill. Alternatively, the required low permeability may be engineered by use of a liner; the various materials that are available for this purpose are discussed in paragraphs 4.54-4.59. At containment sites, attenuation processes will take place almost entirely within the body of the waste. In time these processes will normally reduce the organic strength of the leachate although in some circumstances particularly where difficult wastes are involved, leachate quality may deteriorate (see paragraph 7.15 et seq).

3.11 Containment sites are not generally suitable for the disposal of large volumes of liquid wastes. Ultimately such sites will become saturated and overflow unless the ingress of liquid is sufficiently reduced by either a suitable cover or low permeability cap (see paragraph 6.22 et seq); it should be noted that a low permeability cap can be as important as a liner in 'safeguarding' water quality. Leachate control systems will need to be employed at containment sites, enabling the extraction of leachate for treatment and disposal when necessary. The restoration and aftercare of containment sites calls for particular care in planning and management.

Protection of water quality

General

3.12 Landfill sites must be selected and engineered to ensure the protection of water resources. Fundamental to the protection of water quality is the control of leachate. The basis of this control is an understanding of the water balance of the landfill (see paragraph 3.33 et seq) taking account of liquid inputs, storage capacity and run-off, so that estimates of the amount and timing of leachate generation can be made and the control measures needed to deal with the leachate can be determined. Such estimates, and the effectiveness of the control measures taken, will need to be continually checked by monitoring (see paragraph 3.66 et seq and Figure 3.12).

3.13 Once leachate leaves the boundary of the landfill it has the potential to pollute both surface and groundwaters. The following paragraphs describe briefly the ways in which leachate can enter these systems and precautions that need to be taken.

Surface waters

3.14 Unless water authority consent has been given, the discharge of leachate to surface waters must be prevented. Surface waters can usually be protected by the design and engineering of drainage systems; however, leachate may escape by various routes and hydrogeological investigations are needed to ensure that all significant routes are detected and dealt with in the design. Some of the more commonly found routes are described below.

3.15 Leachate may pass through the base or sides of a landfill to issue at the surface at a lower level and then flow into surface waters. Alternatively sites having a relatively impermeable base and sides will retain leachate so that, if liquid inputs are not controlled, it will accumulate and ultimately overflow or break through. Egress of leachate is a risk particularly with some older mineral excavations where drainage adits may have been constructed. Similarly the use of explosives in the past may have enhanced the fissure network. Old mineral excavations should therefore be examined, the sink point and subsequent liquid issue identified and its significance evaluated.

3.16 Small V-sectioned valleys may provide space for landfilling by constructing a culvert over the natural channel and landfilling on top. However, there is the risk that leachate will gravitate to the culvert, track along its outside and possibly discharge into it. Valley sites should be surveyed to identify seasonal and permanent springs, small streams and water courses and the nature and extent of their catchment must be ascertained to determine the maximum capacity of the culvert or diversion channel. Landfilling over culverts is not regarded as good practice unless a well designed and properly engineered culvert is provided. It is better to divert upstream water around such sites if at all practicable.

3.17 Disused railways cuttings may also be considered as artificial valleys, though it should always by born in mind that railway engineers would have made provision to drain their cuttings and these drains are likely to remain. It must be determined whether drains are present and, if so, whether they should be retained, removed, grouted up or diverted.

3.18 Subsidence basins can be formed as a consequence of differential settlements of the land surface resulting from certain methods of underground extraction of minerals such as coal and salt. In wet weather water may accumulate in such basins. This may be aggravated by subsidence affecting the natural drainage characteristics of the area and result in permanently submerged hollows.

Figure 3.12 LEACHATE MONITORING BOREHOLE

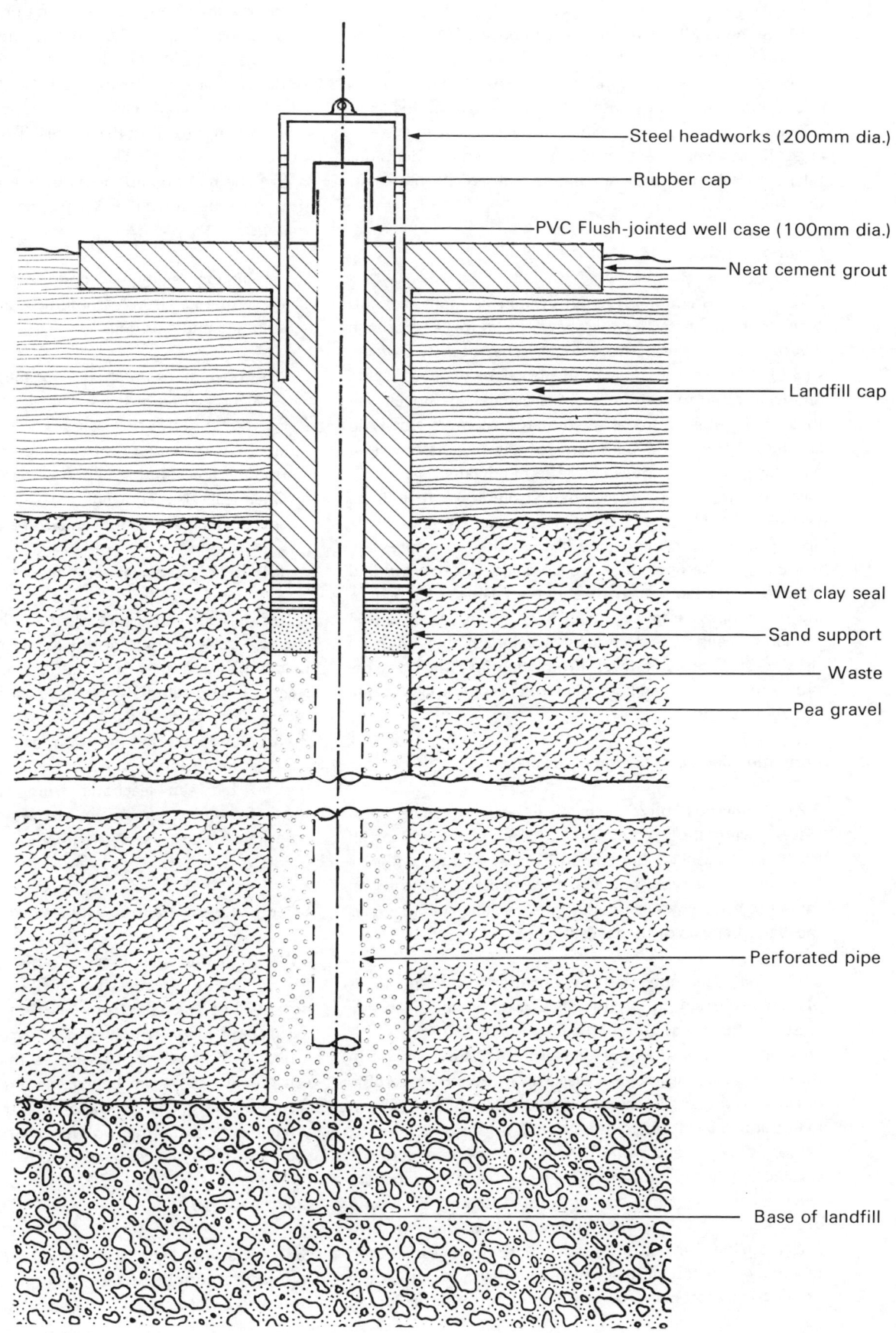

Steel headworks (200mm dia.)

Rubber cap

PVC Flush-jointed well case (100mm dia.)

Neat cement grout

Landfill cap

Wet clay seal

Sand support

Waste

Pea gravel

Perforated pipe

Base of landfill

Alternatively, basins may be dry owing to fractures affording ready egress of water into underlying strata. Such sites would not normally be candidates for a landfill unless they could be adequately engineered to prevent pollution, including any that might occur if further subsidence were to take place.

3.19 Some landfill sites are raised above the level of the surrounding land. In such cases leachate will flow either through the base of the fill material into the underlying deposits or else emerge onto lower-lying surrounding land. It needs to be established whether local drainage systems exist; if so, their extent must be determined and decisions taken on whether they should be utilised, protected or removed.

Groundwaters

3.20 The protection of groundwaters is more difficult than that of surface waters and calls for a thorough investigation of the hydrogeological characteristics of the site to establish the fate of any leachate. At sites where relatively rapid migration occurs, leachate will move away through fissures and coarse sediments, or more slowly in an unsaturated zone consisting of diverse intergranular pathways, before entering the groundwater. At other locations, movement of leachate from the site will be severely restricted but nevertheless some migration must be expected to occur in time. For all sites it is necessary to assess, in the light of hydrogeological information and other relevant factors, the likely effectiveness of attenuating mechanisms (see paragraph 3.27 et seq), both within the landfill and in underlying strata, in reducing pollution risks to an acceptable level. From such investigations the type and extent of engineering works required at the site can be determined.

Formation and composition of leachate

3.21 Numerous physico-chemical and biological processes govern the production and composition of landfill leachates. In general the composition of leachate will be a function of the types and age of waste deposited, the prevailing physico-chemical conditions, the microbiology and the water balance of the landfill.

3.22 The decomposition of putrescible waste takes place by the action of microbes reacting in the waste mass. It occurs, as summarised in Figure 3.73 in which decomposition is shown as taking place in three stages. In the first stage, degradable waste is attacked by aerobic organisms, present in the waste in the presence of oxygen in entrapped air, to form more simple organic compounds, carbon dioxide and water. Heat is generated and the aerobic organisms multiply. The second stage commences when all the oxygen is consumed or displaced by carbon dioxide. Aerobic organisms which thrived when oxygen was available die back. The degradation process is then taken over by organisms which can thrive in either the presence or absence of oxygen. These organisms can break down the large organic molecules present in food, paper and similar waste into more simple compounds such as hydrogen, ammonia, water, carbon dioxide and organic acids. During this stage carbon dioxide concentrations can reach a maximum of 90%, but usually achieve about 50% of the gas generated. In the third and final anaerobic stage, species of methane-forming organisms multiply and break down organic acids to form methane gas and other products. (More details on the process are given in Appendix 3a.) The water soluble degradation products from these biological processes together with other soluble components in the waste will be present in leachate.

3.23 The main components in the leachate from landfill sites may be conveniently grouped into four classes, as follows:

(a) major elements and ions such as calcium, magnesium, iron, sodium, ammonia, carbonate, sulphate and chloride;

(b) trace metals such as manganese, chromium, nickel, lead and cadmium;

(c) a wide variety of organic compounds which are usually measured as Total Organic Carbon (TOC) or Chemical Oxygen Demand (COD); individual organic species such as phenol can also be of concern;

(d) microbiological components.

All household waste and most industrial waste will give rise to leachate. Household waste is reasonably consistent in composition over all landfill sites, as is the resulting leachate; typical figures for leachate composition for recent and aged household waste are shown in Table 3.23A. Leachate composition at sites taking industrial waste is much more variable; some data providing a comparison between leachates from household waste and those from sites taking industrial and household wastes co-disposal sites: (see paragraph 7.7) are given in Table 3.23B.

Effects of water pollution by leachate

3.24 In earlier sections of this chapter and indeed throughout this report the need to reduce the polluting potential of leachate from landfill sites is stressed. Good design and management, and in particular the full use of attenuation processes, can ensure that water resources are not put at risk. Nevertheless, the potential to affect water resources severely exists and should be understood. The effects of substantial pollution of surface waters and groundwater by leachate are briefly described below.

Surface waters

3.25 The input of high strength organic leachate and inorganic solutions of metals in a reduced state of oxidation into a watercourse will deplete the oxygen content of

Table 3.23A Typical composition of leachates from recent and aged domestic wastes at various stages of decomposition (all results in mg/l except pH-value)

Determinand	Leachate from recent wastes	Leachate from aged wastes
pH-value	6.2	7.5
COD (Chemical Oxygen Demand)	23 800	1 160
BOD (Biochemical Oxygen Demand)	11 900	260
TOC (Total Organic Carbon)	8 000	465
Fatty acids (as C)	5 688	5
Ammoniacal-N	790	370
Oxidised-N	3	1
o-Phosphate	0.73	1.4
Chloride	1 315	2 080
Sodium (Na)	960	1 300
Magnesium (Mg)	252	185
Potassium (K)	780	590
Calcium (Ca)	1 820	250
Manganese (Mn)	27	2.1
Iron (Fe)	540	23
Nickel (Ni)	0.6	0.1
Copper (Cu)	0.12	0.3
Zinc (Zn)	21.5	0.4
Lead (Pb)	8.4	0.14

the water and ultimately will result in the extinction of all oxygen-dependent life. If the leachate contains non-biodegradable. organic compounds these will persist for some time. When such compounds are assimilated into food chains they may adversely affect aquatic species. Although the effect on aquatic fauna of individual compounds can be assessed or predicted with some confidence the combined effects of several compounds cannot. In addition temperature, pH and dissolved oxygen concentration all have an influence on the degree of toxic effects on a particular aquatic species. For many pollutants, no single concentration can be given which is universally applicable for all aquatic environments.

Groundwater

3.26 The effects of high strength organic leachate on groundwater will persist for a long time due to the limited amount of dissolved oxygen available and the low rates of dispersion. Once the groundwater has become polluted it may be unsuitable as a source of potable water supply for many years. The nature of the strata and groundwater flow rates will control the extent of the pollution plume and its movement down the groundwater gradient away from the waste disposal site. Some constituents in leach-

Table 3.23B Composition of leachates from solid wastes in landfills (all results in mg/l except pH-value)

	Household waste	Pitsea (UK) (43% industrial)	Rainham (UK)* (Industrial/household)	Granmo (Norway) (66% industrial)	Cedar Hills (USA) (Industrial/household)
pH-value	5.8–7.5	8.0–8.5	6.9–8.0	6.8	5.4
COD	100–62 400	850–1 350		470	38 800
BOD	2–38 000	80–250		320	24 500
TOC	20–19 000	200–650	77–10 000	100	
Volatile acids (C$_1$–C$_6$)	ND–3 700	20	600–10 000	10	7 100
Ammoniacal-N	5–1 000	200–600	90–1 700	120	
Organic-N	ND–770	5–20		62	
Nitrate-N	0.5–5			0.04	
Nitrite-N	0.2–2	0.10–10	8.0		
o-Phosphate	0.02–3	0.20		0.6 (Total)	11.3 (Total)
Chloride	100–3 000	3 400	400–13 00	680	
Sulphate	60–460	340	150–1 100	30	
Sodium (Na)	40–2 800	2 185	2 000	462	
Potassium (K)	20–2 050	888	50–125	200	
Magnesium (Mg)	10–480	214		66	
Calcium (Ca)	1.0–165	88		188	
Chromium (Cr)	0.05–1.0	0.05	0.5	0.02	1.05
Manganese (Mn)	0.3–250	0.5			
Iron (Fe)	0.1–2 050	10	0.6–1 000	70	810
Nickel (Ni)	0.05–1.70	0.04	0.5	0.1	1.20
Copper (Co)	0.01–0.15	0.09	0.5	0.09	1.30
Zinc (Zn)	0.05–130	0.16	1.0–10	0.06	155
Cadmium (Cd)	0.005–0.01	0.02		0.0005	0.03
Lead (Pb)	0.05–0.60	0.10	0.5	0.004	1.40
Monohydric phenols		0.01	ND–2.0		
Total cyanide		0.01	0.09–0.52		
Organochlorine pesticides		0.01			
Organophosphorous pesticides		0.05			
PCBs		0.05			

* Samples obtained from boreholes within the fill.

ate are chemically stable while others are not amenable to attenuation by physical processes in the groundwater environment. The identification and quantification of a pollution plume is difficult to achieve unless a large number of boreholes are constructed and groundwater samples abstracted and analysed at regular intervals.

Attenuation

3.27 When waste is landfilled, various processes acting within the landfill, including dilution and dispersion, tend to reduce the polluting potential of the leachate. Leachate quality will also be improved, by dilution and other processes, in situations where it migrates from a site through surrounding geological strata. The action of these processes as a whole, which may be physical, chemical or biological in character, is termed 'attenuation' (see Figure 3.27). Attenuation processes vary according to the nature and quantities of the wastes deposited. The degree of attenuation that may be achieved outside the landfill depends on both the geochemistry of the strata and prevailing hydrogeological conditions; in particular, processes operating within an unsaturated zone may result in significant attenuation of polluting species.

3.28 Attenuation may result in either the effective removal of a particular polluting component in the leachate, or its delayed release into the environment. The latter is important since a relatively prolonged release at low concentration is often preferable to a rapid release at a high concentration. The mechanisms of individual attenuation processes are often as yet insufficiently understood. However, experience indicates that attenuation typically results in concentrations of polluting components in the leachate several orders of magnitude lower than those present in the waste. The principal attenuation processes are summarised in Table 3.28 and are described briefly in the following paragraphs.

Table 3.28: **Attenuation processes**

Physical	Absorption, adsorption, filtration, dilution dispersion.
Chemical	Acid-base interactions, oxidation, reduction, precipitation, co-precipitation, ion-exchange, complex ion formation.
Biological	Aerobic and anaerobic microbial degradation.

Physical processes

3.29 Physical sorption processes are significant in the attenuation of polluting species. For example, landfilled solid waste may absorb leachate, thus preventing its rapid migration and thereby giving time for attenuation processes to act. The quantities, proportions and solubilities of wastes deposited affect leaching. In comparison with well dispersed deposits, large discrete deposits will

generally leach more slowly, but the concentration of each potentially polluting component in leachate close to the deposit is higher. The depth of waste, its density and the extent to which fissures and channelling allows rapid leachate movement through the landfill is particularly important. Rapid movement of leachate gives little time for physical attenuation processes to act. In such situations recirculation of leachate may be beneficial.

3.30 Dilution is also an important attenuation process. One aspect of this is the dispersal of an industrial waste throughout a landfill which should lead to lower concentrations of the polluting species in the leachate albeit in a larger volume. Further dilution will occur if the leachate reaches ground or surface waters, the oxygen content of which can promote further biodegradation.

Chemical processes

3.31 Chemical attenuation relies on the polluting species being immobilised or chemically changed. The solubility of most species tends to increase as pH decreases. Oxidation — reduction reactions can also occur. Precipitation and co-precipitation may decrease the mobility of potential pollutants. In addition the complexing of metallic ions may also be important. It is possible to enhance these processes by the mode of operation of the landfill for example by recirculation, layering of wastes, promoting the formation of insoluble chemical compounds using selected wastes, and the like.

Biological processes

3.32 At present, although, not so well understood as other processes, biological mechanisms are likely to have an important role to play. The microbial population in a landfill is related to the type, age, depth, compaction, moisture content, and density of the waste. Both aerobic and anaerobic microbiological degradation can occur, anaerobic conditions being likely to predominate in a well-compacted site. Microbial activity is responsible for the generation of methane gas (see paragraph 3.22).

Water balance calculations

3.33 The importance of understanding the water balance of a landfill is referred to in paragraph 3.12. Leachate generation should be minimised. The effects of the infiltration of rainfall make it imperative that both ground and surface water ingress into the landfill are controlled (see Figure 3.33). A useful aid in calculating and assessing the likely rate and extent of leachate production is to apply water balance equations. The main factors contributing to the water balance of a landfill site are:

(a) water input including effective rainfall (precipitation minus run-off and evapotranspiration), surface and groundwater infiltration and liquid waste disposal,

(b) surface area,

Figure 3.27   THE VARIOUS FACTORS WHICH CAN CONTRIBUTE TO THE ATTENUATION OF CHEMICAL SPECIES BY LANDFILLS AND SURROUNDING GEOLOGICAL STRATA

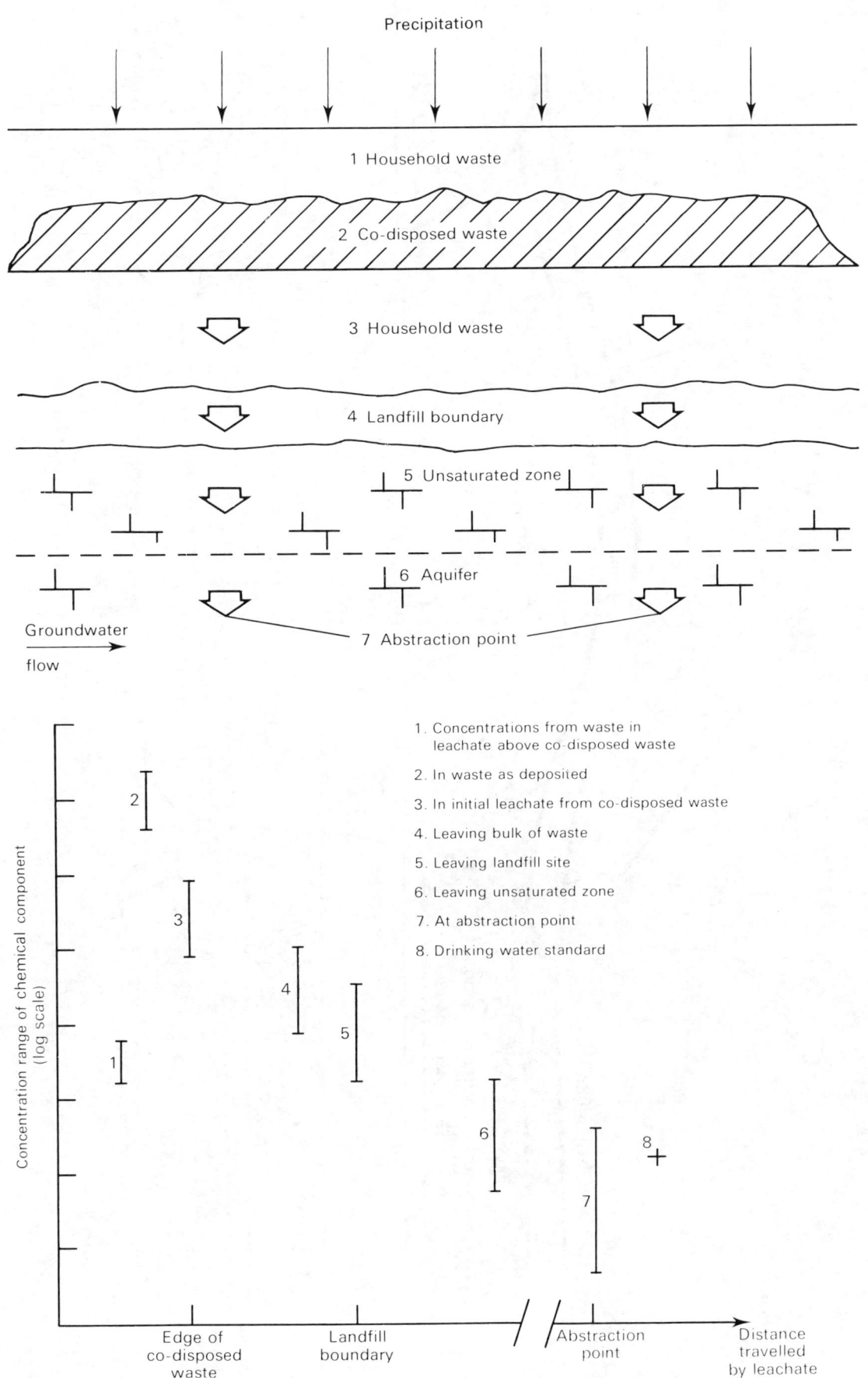

Precipitation

1 Household waste

2 Co-disposed waste

3 Household waste

4 Landfill boundary

5 Unsaturated zone

6 Aquifer

7 Abstraction point

Groundwater flow

Concentration range of chemical component (log scale)

1. Concentrations from waste in leachate above co-disposed waste
2. In waste as deposited
3. In initial leachate from co-disposed waste
4. Leaving bulk of waste
5. Leaving landfill site
6. Leaving unsaturated zone
7. At abstraction point
8. Drinking water standard

Edge of co-disposed waste

Landfill boundary

Abstraction point

Distance travelled by leachate

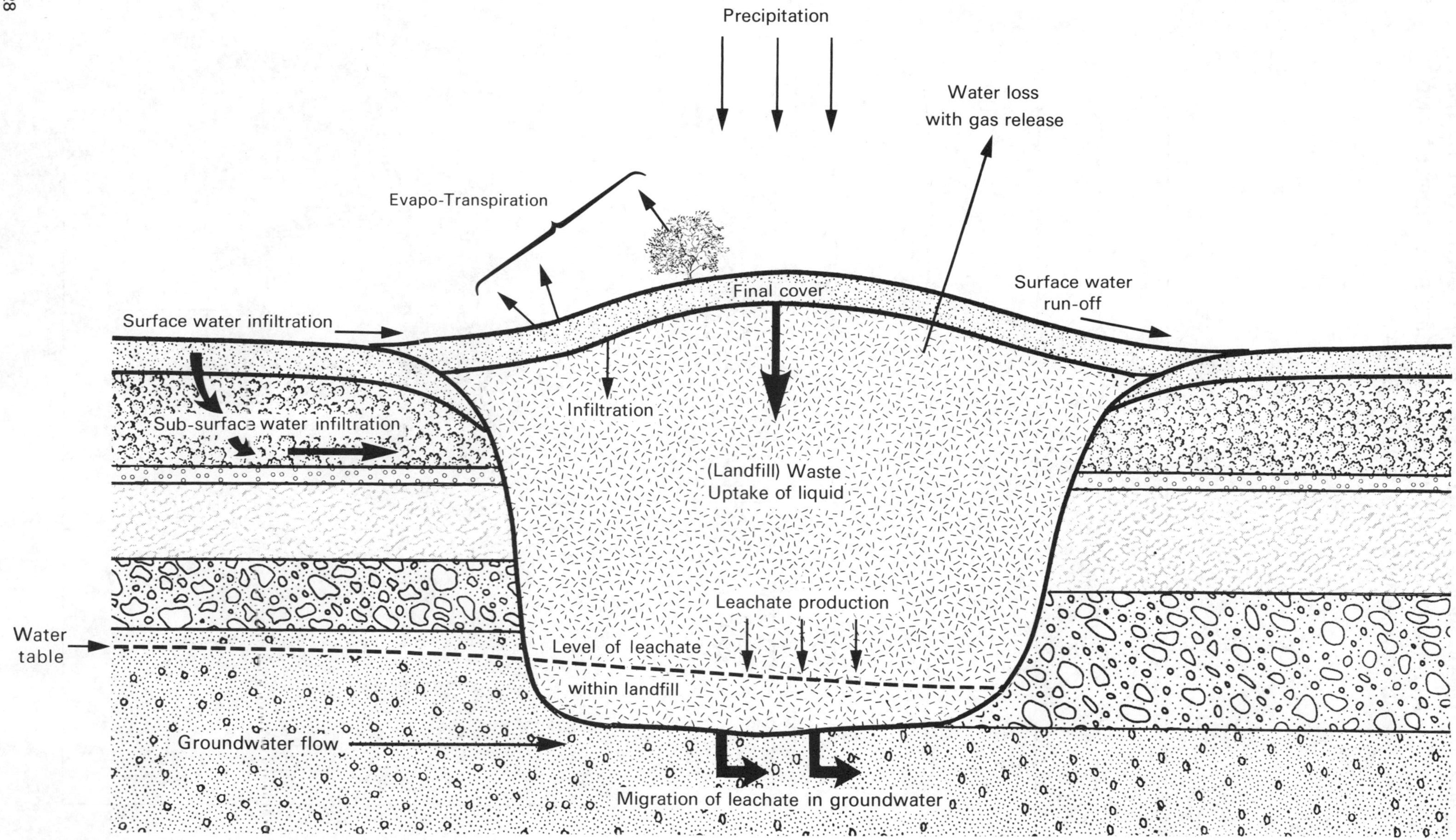

Figure 3.33 LANDFILL WATER BALANCE

Precipitation

Water loss with gas release

Surface water run-off

Evapo-Transpiration

Final cover

Infiltration

(Landfill) Waste Uptake of liquid

Leachate production

Level of leachate within landfill

Migration of leachate in groundwater

Surface water infiltration

Sub-surface water infiltration

Groundwater flow

Water table

28

(c) nature of wastes,

(d) site geology, and

(e) surface liquid storage eg leachate balancing lagoons.

3.34 In estimating the amount of leachate likely to be generated each year from operational areas the following water balance equation may be used:

$$L_o = I - E - aW \qquad (i)$$

where

L_o = free leachate retained at the site (equivalent to leachate production minus leachate leaving the site) (m^3/annum)

I = total liquid input (precipitation plus liquid waste plus any surface or ground water inflow) (m^3/annum)

E = evapotranspirative losses (evaporation plus minimal transpiration) (m^3/annum)

a = absorptive capacity of the waste (m^3/tonne of waste as received)

W = weight of waste deposited (tonnes/annum)

Good landfill practice normally requires that the site is operated so that L_o is always negative or zero. A positive value for L_o implies leachate build-up in the site.

3.35 The equation requires modification to cover the position when final restoration has taken place. There may then still be some further absorptive capacity for liquid within the waste, (U) and an additional quantity (R) must be introduced to allow for surface water runoff. The revised water balance equation is then:

$$L_r = I - E - R - U \qquad (ii)$$

where; L_r = leachate retained in the site after restoration.

For sites having a low permeability cap (see paragraph 6.22), R will be very high. The term U may be calculated by summing the values of L_o using equation (i), during the operation of the site. The major factors controlling leachate production are described below, with reference to the terms and in the water balance equations.

Total liquid input (I)

3.36 Rainfall, which in the UK ranges between 0.55 and 2.0 metres per year is normally the largest contributor of liquid to a landfill site. As far as is practicable surface and groundwater infiltration into a site should be eliminated. Any residual input however, should be estimated and included. Methods for controlling such infiltration are dis-

cussed in Chapter 4. Some sites are licensed to accept liquid wastes and for the purposes of water balance calculations these should be included as part of the total liquid input. Spraying leachate generated on site back over fresh waste or the completed site is a relatively common practice. This leachate will already have been accounted for in water balance calculation for the operational landfill. However if it is sprayed over new areas of the site or over completed parts it should be included as part of the input term in equations (i) and (ii) as appropriate. Although input data to the water balance equation above (i) have been assumed to be on an annual basis it may be prudent to make the calculations on a seasonal basis in areas of the UK where wide seasonal variations in precipitation are likely.

Evapotranspiration (E)

3.37 Wind, temperature, humidity and atmospheric pressure influence the rates of evaporation/evapotranspiration, and are reflected in the value of term E in equations (i) and (ii) above. However, whilst climatic conditions vary from one area of the UK to another, the resulting range in the value of E is likely to be much less than the differences that occur between values at operational and completed landfills. Healthy vegetation on completed landfills increases evapotranspiration and also binds the soil, thus minimising erosion. However, the effectiveness of final cover to limit infiltration and encourage surface water run-off may be impaired by penetration by roots.

3.38 The effective rainfall at a site is given by the actual rainfall minus E. The Meteorological Office can provide effective rainfall figures based on climatic measurements at selected sites. The amount of rainfall that infiltrates a landfill will be the effective rainfall less any run-off. Experimental data suggest that infiltration rates through typical daily cover soils range from 20% in summer months to 100% during the winter. Such high infiltration rates, which are apparently independent of the quality of typical daily cover used, indicate the importance of limiting operational areas and excluding other sources of liquid ingress. Similar conditions apply to intermediate cover; evidence suggests that the infiltration rate of precipitation is not significantly altered by the type or quantity of intermediate cover used. If, during infilling, surfaces are formed with slopes greater than 10 per cent (1 in 10) then some run-off can be expected but this will not substantially reduce overall infiltration, unless collection and removal of any run-off water is provided. For these reasons it is most unlikely that run-off will be effective at an operational site and it has therefore not been included in equation (i).

3.39 The position is very difference for completed landfills where, with the use of contouring and the incorporation of drainage systems (and, at containment sites, the use of a low permeability cap — see paragraph 6.22) a substantial proportion of precipitation can be diverted away from the site as run-off. (Hence the introduction of the factor (R) in equation (ii) above.) Typical infiltration values through completed sites are 25-30% of annual effective rainfall. In addition to quantity, depth, compaction and surface profile of cover material, the effectiveness of run-off will also be affected by the frequency, intensity and duration of rainfall events.

Absorptive capacity of solid waste (a)

3.40 While take-up of liquid by sorption can effectively delay the generation of leachate this on its own will not prevent the build-up of leachate in a landfill. The sorption of liquid by waste is a significant component in the water balance calculation.

3.41 There are two main mechanisms for moisture retention within waste. First, liquid can be absorbed into the waste and held by capillary attraction within the micro structure of waste particles. Second, is the storage of free water in voids within the waste. Typical values for overall voidage are between 20 and 35%. Intermediate cover materials or highly compacted wastes can also give rise to localised areas of saturation known as perched water tables. Thus water retention values are strongly dependent on both the density of the waste and the presence or absence of voids or of impermeable barriers which will inhibit the continuous downward percolation of liquid.

3.42 At many landfill sites waste densities of 0.7 to 0.8 tonnes/m^3 of waste as received are achieved and at such densities it is likely that about 0.1 to 0.2 m^3 of added liquid per cubic metre of waste as received can be absorbed before substantial leachate generation commences. However, at higher compaction densities absorptive values will fall. For example, at placement densities in excess of 1.0 t/m^3 the absorptive capacity may fall to as low as 0.02-0.03 m^3 liquid/m^3 of as received waste.

Waste (W)

3.43 Finally a value for the annual quantity of waste to be deposited (W) is required to calculate an overall water balance for the site. Generally speaking the higher the rate of waste input the less likely is it that leachate will be generated during the operational phase of a site. This can mean that the absorptive capacity of the waste will not be utilised and landfill gas generation as a consequence of biodegradation cannot be promoted in the absence of sufficient water input.

Leachate treatment

General

3.44 Leachate accumulated within a landfill or collected from drainage systems designed to protect surface waters may need to be disposed of to the environment, and treatment may first be needed to reduce its polluting potential. At present the dominant method of disposal is the discharge of leachate to sewer, land, water course or

tidal water under consent conditions issued by the water authority concerned. Where discharge is to a sewer, treatment of the leachate takes place at the sewage works (where volumes of leachate generated are low, tankering to a sewage works may be the most appropriate method of disposal). This is one form of off-site treatment, ie treatment at a facility not directly associated with the landfill operation. Other options for off-site treatment are: treatment at specialised hazardous or toxic waste centres and sea outfall disposal. So far, in the UK, on-site leachate treatment at a landfill is still only at a pilot scale of development. The quality of effluent produced is not generally suitable for discharge to inland surface waters. The on-site treatment of effluents is, however, carried out in a number of industries in order to meet consent conditions set by water authorities. The processes and technologies available for leachate treatment are described in this section (see also paragraph 7.17).

3.45 The volume and strength of leachate produced at landfill sites is subject to large seasonal variations. Wide fluctuations in flow and concentration can be minimised by balancing leachate flow, either by storage within already deposited waste or by using a lagoon, so reducing the required treatment capacity by removing the peak loadings. However, concentrations of components in leachate also change with its age. Treatment strategies must therefore adapt to changes in leachate volumes and strengths both during the filling stage of the landfill and after its completion. Leachate, particularly that from recently emplaced wastes, contains high concentrations of readily biodegradable material (principally organic acids), which are amenable to biological treatment. Leachates from wastes which have been deposited for a longer time are generally lower in organic content, are less readily biodegradable and may contain relatively high concentrations of ammonia or iron. Thus leachate from aged waste may require a combination of processes for effective treatment.

3.46 The method and degree of leachate treatment necessary will depend upon the consent criteria set by the water authority. Water authorities, responsible for setting consent conditions, will be concerned with such parameters as organic loading, metals, sulphides, chlorinated hydrocarbons, ammonia, pesticides and herbicides. Consent criteria laid down for discharges from landfills will reflect the type of waste deposited, any expected variation in flow and strength of toxic components in the leachate, and the nature and point of discharge (for example, whether sewer or stream). Consent criteria will generally be site specific.

On-site treatment technologies

3.47 Technologies for on-site treatment of leachates, including those containing toxic or hazardous components, include:

(a) biological anaerobic or aerobic systems

Figure 3.42 GENERAL DIAGRAM OF RELATIONSHIP OF MOISTURE CONTENT VS WASTE DENSITY

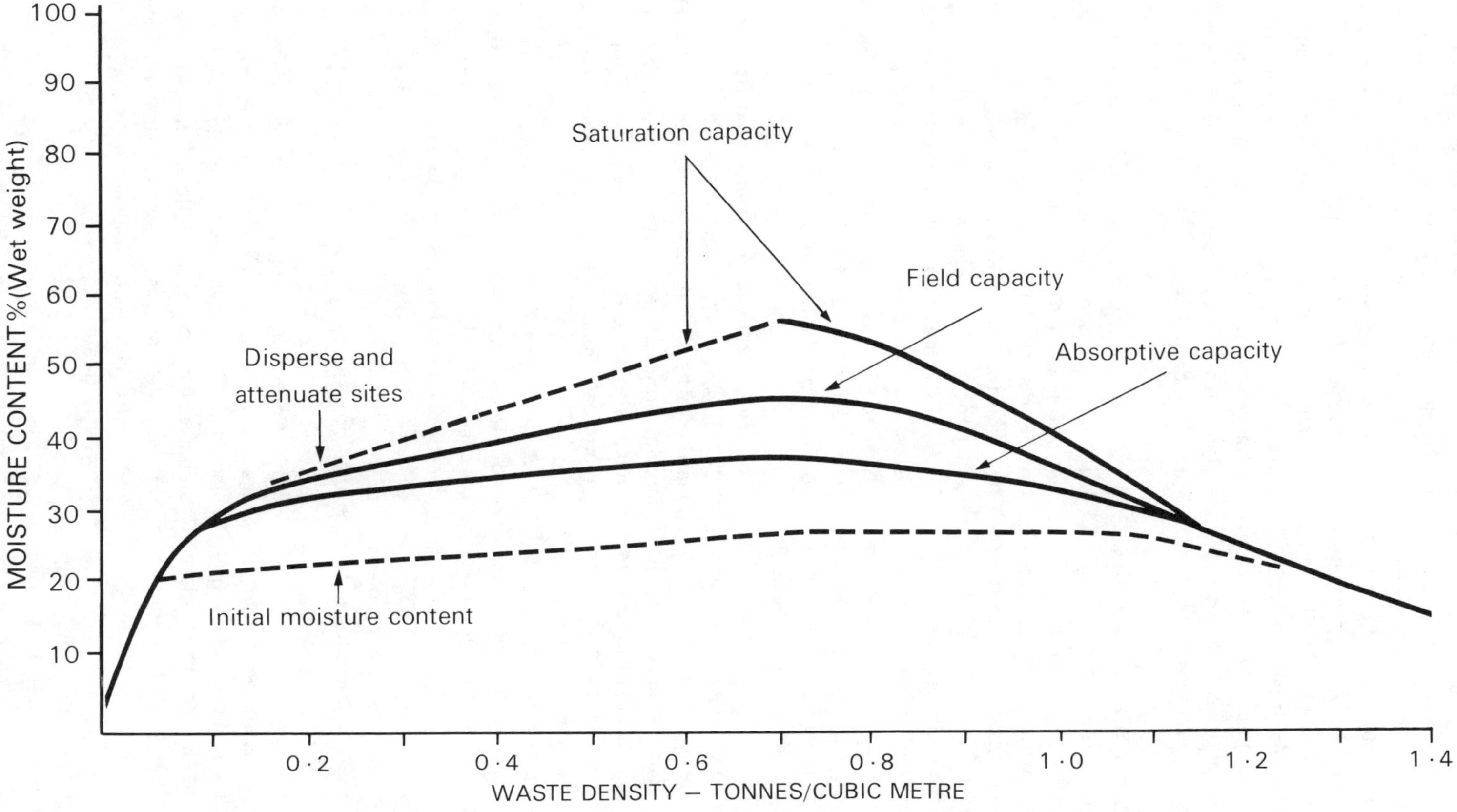

(b) chemical oxidation and reduction

(c) precipitation

(d) air stripping

(e) carbon adsorption.

Selection of treatment process should be based on leachate quality, laboratory evaluation studies and, where possible, on pilot scale studies. Where no leachate data exist, such as at the design stage for a new landfill, an alternative approach is required. In this situation leachate quality may be predicted from leachate generation calculations, experience at other sites and leaching tests for typical industrial wastes expected to be received.

3.48 Biological treatment processes may be adversely affected by toxins occurring in both household and industrial waste but various factors may act to moderate such effects. The micro-organisms which form the basis of biological treatment may prove resistant to transitory shock discharges of toxic substances. Even if the microbial population is severely depleted, the situation can usually be retrieved since the regeneration of some micro-organisms is rapid. Finally, many biological populations have the capacity to acclimatise to the point of actually utilizing many toxins. Biological treatment of leachate can be accomplished by both aerobic and anaerobic processes.

3.49 In addition to their ability to oxidise certain toxic organic substances, aerobic processes can also acclimatise to the presence of heavy metal ions. In such cases the heavy metals are not oxidised by biological processes but are adsorbed or absorbed by the biological floc. Even though biological treatment processes can be effective in removing many toxic substances, pre-treatment may be necessary if they are present at abnormally high concentrations. The compounds which may adversely affect biological treatment include:

(a) Metals — If inhibition by metals is a problem simple pre-treatment with lime to precipitate the metals as hydroxides may be necessary.

(b) Carbon compounds — Very high concentrations of degradable organic compounds can be oxidised biologically by acclimatised systems. Low concentrations of chlorinated solvents will rapidly volatilise during aerobic treatment.

(c) Ammonia — High concentrations of ammonia (up to 2,500 mg/l) appear to be tolerated both by aerobic and anaerobic systems but the biological treatment rate will be reduced at high ammonia concentrations.

(d) Chloride — Chloride can be tolerated at relatively high concentrations (up to 20,000 mg/l) by aerobic processes. Anaerobic processes appear slightly more sensitive with 10,000 mg/l inhibiting gas production.

(e) Sulphide — Anaerobic digestion processes usually tolerate up to 200 mg/l as soluble sulphide and little effect has been observed at concentrations up to 400 mg/l. In contrast aerobic processes can satisfactorily treat 1,000 mg/l without impairing performance.

3.50 Biological treatment process plants used include anaerobic lagoons or reactors and, as aerobic systems, aerated lagoons and activated sludge systems. Physico-chemical processes considered to be of most use on landfill leachates include air stripping, pH adjustment, chemical precipitation, oxidation and reduction. Advanced techniques such as carbon adsorption, ion exchange and solvent extraction may be necessary for certain leachates. These treatment technologies are further described below.

3.51 Aerobic treatment systems generally use aerated lagoons or the activated sludge process. In both cases the leachate is aerated in the presence of an active biological floc. For aerated lagoons retention periods in excess of five days may be required to achieve greater than 95 per cent removal of the chemical oxygen demand (COD) from leachate. To maintain biological activity within an aerobic system, a supply of essential nutrients is necessary. Leachate is normally deficient in phosphorus and occasionally nitrogen, and therefore addition of these nutrients may be required.

3.52 Activated sludge systems differ from aerated lagoons by using sludge recirculation and a separate clarifier. The overall oxygen demand is thereby reduced and a more active biological community can be maintained within the reactor allowing the hydraulic retention time to be reduced and treatment rate increased.

3.53 Aerobic systems can be simple to design, construct and operate using lined basins fitted with surface aerators or coarse bubble aeration plant. Biological sludges produced in aeration basins can be landfilled but are difficult to dewater.

Anaerobic reactors

3.54 Anaerobic biological processes offer several advantages over aerobic ones when used for the treatment of leachate. The advantages include the generation of methane gas and a much lower rate of production of sludges. Also, an aerobic system does not need aeration equipment. The main disadvantage is that ammonia is not effectively treated.

3.55 In an anaerobic treatment system, complex organic molecules present in the leachate are fermented by bacteria to carboxylic acids, which are then partially converted to methane and carbon dioxide by methanogenic bacteria. Most of the development work carried out using anaerobic biological processes has been on a small scale using mixed digesters or anaerobic filters. Although filters have been found to be more efficient than mixed digesters both systems are generally considered to be unsuitable for use with leachates for on-site treatment.

3.56 Physical and chemical processes have been investigated as an addition to, or replacement for, aerobic or anaerobic biological treatment of leachate. Chemical methods used to treat leachate have been the addition of selected chemicals to precipitate, oxidise or reduce both inorganic or organic fractions. Precipitation and coagulation treatment using lime, ferric chloride or alum has been shown to have little effect on the removal of organic matter. Ion exchange processes have been found to be effective but are expensive. Removal of colour, suspended solids and heavy metals however, appears to be effective.

3.57 Attempts to remove organic material from leachate using chemical oxidants have been made. Hydrogen peroxide, ozone and, less commonly, calcium hypochlorite and potassium permanganate have been evaluated. Excellent removal of colour and iron has been achieved. To obtain any significant reductions in COD high dose rates were required. The amelioration of certain odours, such as that due to hydrogen sulphide, has been successfully carried out by treatment with hydrogen peroxide. It has also been used successfully to oxidise sulphide in leachate to sulphate before discharge to sewer.

3.58 Leachate often contains high concentrations of nitrogen compounds, principally as ammoniacal nitrogen. Removal of ammoniacal nitrogen has been evaluated at both full-scale and pilot-plant scale. For air stripping of ammoniacal nitrogen it is converted to a gaseous form (ammonia) at pH-values 10.5-11.5 and removed by contact with air either in a stripping tower or pond system. Tower stripping is impractical at low termperatures because of icing problems. The use of shallow tanks and coarse bubble aerators has been successful and is recommended.

Other on-site treatment options

3.59 Two further approaches to the treatment and disposal of leachates are described below. The first of them, land irrigation, can be carried out on-site only if sufficient land is available.

3.60 Spraying of leachate onto land (as opposed to landfill sites) can result in a significant reduction in its volume due to evapotranspiration. Additionally, as the leachate percolates through vegetated soils opportunities are provided for microbial degradation of organic components, removal of inorganic ions by precipitation or ion exchange and the possibility for rapid uptake of constituents such as ammonia by plants.

3.61 Evaporative losses may be maximised by using standard agricultural mist sprays. Intermittent spraying throughout each day will provide more effective evaporation than a single daily application. Transpiration by vegetation will account for a substantial proportion of

total loss. The possibility of spreading harmful pathogens by spraying leachate needs to be considered but evidence to date suggests that this is not a problem provided the operation is properly managed.

3.62 Little information is available on the long-term effects of continual spraying of leachate onto land. The spraying of leachates containing metals or persistent organic compounds is not to be recommended because of their accumulation in soils and plant material. In this respect reference should be made to publications on the application of sewage sludge to land and contaminated land trigger concentrations.

3.63 This form of leachate treatment requires large areas of land. If the treatment gives rise to run-off away from the landfill a consent to discharge will be required from the water authority. Land irrigation as a treatment process may require planning permission and a disposal license separate from those granted for the landfill site operation.

Recirculation through the landfill

3.64 Recirculation of leachate through solid wastes, particularly where the biochemical activity of the waste has not been exhausted, potentially offers advantages both in reducing the volume of liquid by evaporation, and reducing its strength. In addition it has been reported that increasing the moisture content of the fill gives rise to a more rapid stabilisation of wastes, and enhances gas production.

3.65 Although it is being utilised extensively in practice, most of the available data on leachate recirculation has been from pilot scale studies. The major benefit shown has been the production of a low strength leachate in a relatively short period of time. The leachates were similar in composition to those produced by wastes at least five years old. The solid wastes also appear to have degraded and stabilised more rapidly. Like land irrigation, recirculation will be most effective in the summer months when ambient temperatures and consequent evaporative losses are high and leachate production is at a minimum.

Leachate monitoring

3.66 A theme running through this chapter is that the landfill is a complex reactor where physical, chemical and biological processes act to transform potentially polluting wastes into environmentally acceptable and beneficial deposits. Because of the complexity of these processes and their potential environmental effects, monitoring is needed to confirm that the landfill is behaving in the ways predicted when the site was first planned and licensed and to provide information needed for management decisions. In this respect landfill is no different from most industrial processes that rely on complex chemical or biological reactions. There is, however, one difference; while closure of an industrial plant usually means that the need to monitor ceases, this is rarely the case for a landfill.

Processes acting within or close to the landfill can continue for many years after the last deposit of waste. Therefore, monitoring should also continue beyond cessation of operations (see paragraph 6.102). Proper planning of the total landfill operation, including aftercare, should overcome this problem.

3.67 Leachate monitoring plays a central role in the management of all landfills. In the first place, data on leachate volume and composition are essential to the proper control of leachate so as to ensure the protection of water quality. While the formation of leachate is almost inevitable in the UK climate, by proper management its uncontrolled generation within a landfill is not. Modern landfilling therefore usually requires some form of leachate treatment. Such treatment can be successful only if the leachate conforms to predicted quality and treatment plant design perameters. Even where leachate can be simply pumped to sewer, its volume and quality needs to be controlled to satisfy water authority consent conditions. Monitoring of leachate volume is therefore linked to monitoring of its chemical composition. A knowledge of the chemical make-up of a leachate is necessary for the effective working of most leachate treatment schemes. In addition to providing adequate storage capacity for leachate and its management, the monitoring data also provides the means of assessing the general behaviour of a landfill. For example, knowledge of the chemical composition of a leachate is required to confirm that attenuation processes within the landfill are proceeding as expected. Changes in chemical composition of leachate can act as a warning system and help identify problems caused, for example, by overloading one part of the site with a particular waste. Remedial action can thus be taken. Bearing in mind the long life of a modern landfill, during which time significant changes could take place in landfill practices and in the character of the wastes deposited, monitoring is necessary to ensure that the measures taken for environmental protection remain effective.

3.68 Leachate monitoring should not be confined to the landfill itself, but should also take place outside the landfill boundary. Two types of landfill have been identified. One relies on attenuation mechanisms acting reasonably rapidly within and near to the landfill in reducing the polluting components of leachate to acceptable levels. The other relies on containing leachate within the landfill for some appreciable time so that attenuation processes have time to act. In both cases monitoring is imperative; in the first, to confirm that attenuation processes are proceeding in the manner predicted, and, in the second, to confirm the integrity of the site, particularly if artificial liners are employed.

3.69 Any monitoring scheme must provide detailed information on the development of leachate within, and beyond, a landfill. The scheme will be site specific; it should be drawn up at the site investigation stage and implemented, as far as practicable, as the first stage in the site preparation (see paragraph 4.51). It cannot be stressed too strongly that it is vital that background readings are obtained before landfilling operations commence against which the results of later monitoring may be judged. Without such background readings it is very difficult to determine what, if any, effect a landfill has on the environment.

3.70 Monitoring of a landfill and its environs is not solely the responsibility of the landfill operator though he must bear the major share. The disposal authority will also wish to undertake some monitoring to ensure that the conditions specified in the disposal licence are being adhered to and remain appropriate. Similarly, in exercising its water pollution prevention duties, the water authority will, no doubt, wish to undertake some monitoring. With three parties involved, there is a need to ensure that duplication of effort does not take place and that an appropriate monitoring scheme is undertaken. All parties should therefore get together early in the site preparation stage to agree the monitoring scheme and particularly who is to be responsible for what. Analyses undertaken by the three parties should be on a consistent basis to facilitate the comparison and pooling of results.

Landfill gas (see also Annex 2)

Introduction

3.71 During the decomposition of the organic materials by biological activity in the waste, various gases and vapours known collectively as landfill gas are evolved. The gas poses hazards, in particular through the risk of fires or explosions, and appropriate precautions must be taken in the design and operation of landfill sites to minimise these risks. In some circumstances gas may be produced in such quantities that its extraction as a fuel becomes worthwhile. This section provides a brief account of the generation and composition of landfill gas and the environmental problems it creates, and of various aspects of site design that relate to the control of gas migration through the site. The considerations that apply to the use of landfill gas as a fuel are also outlined. However, the control and use of landfill gas raises many issues that requires special consideration including: legal aspects; the health and safety of personnel; the selection and design of plant and equipment; and monitoring requirements. A detailed discussion of these issues is presented in Annex 2.

3.72 Several factors influence the rate at which landfill gas is generated. The moisture content of the waste is important as increased moisture enhances decomposition and hence gas generation. The decomposition process is complex (see Annex 2 and Figure 3.72). Apart from the uppermost metre or so, landfills are generally anaerobic. Optimum conditions for anaerobic digestion normally occur in unsaturated wastes with greater than 40 per cent moisture content. Methane forming bacteria can survive only within a narrow pH range, the optimum range for

Figure 3.72   DECOMPOSITION OF MATERIALS OCCURRING IN HOUSEHOLD WASTE

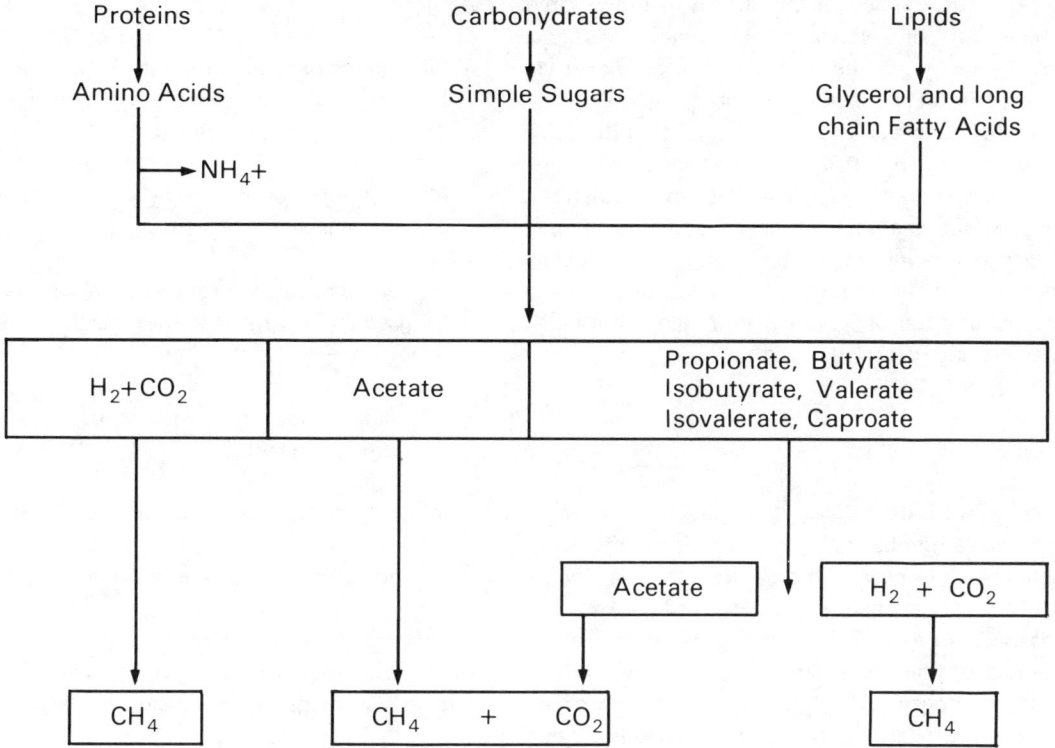

Figure 3.73   (b) TYPICAL PRODUCTION PATTERN FOR LANDFILL GAS
(Adapted from Farquhar and Rovers, 1973)

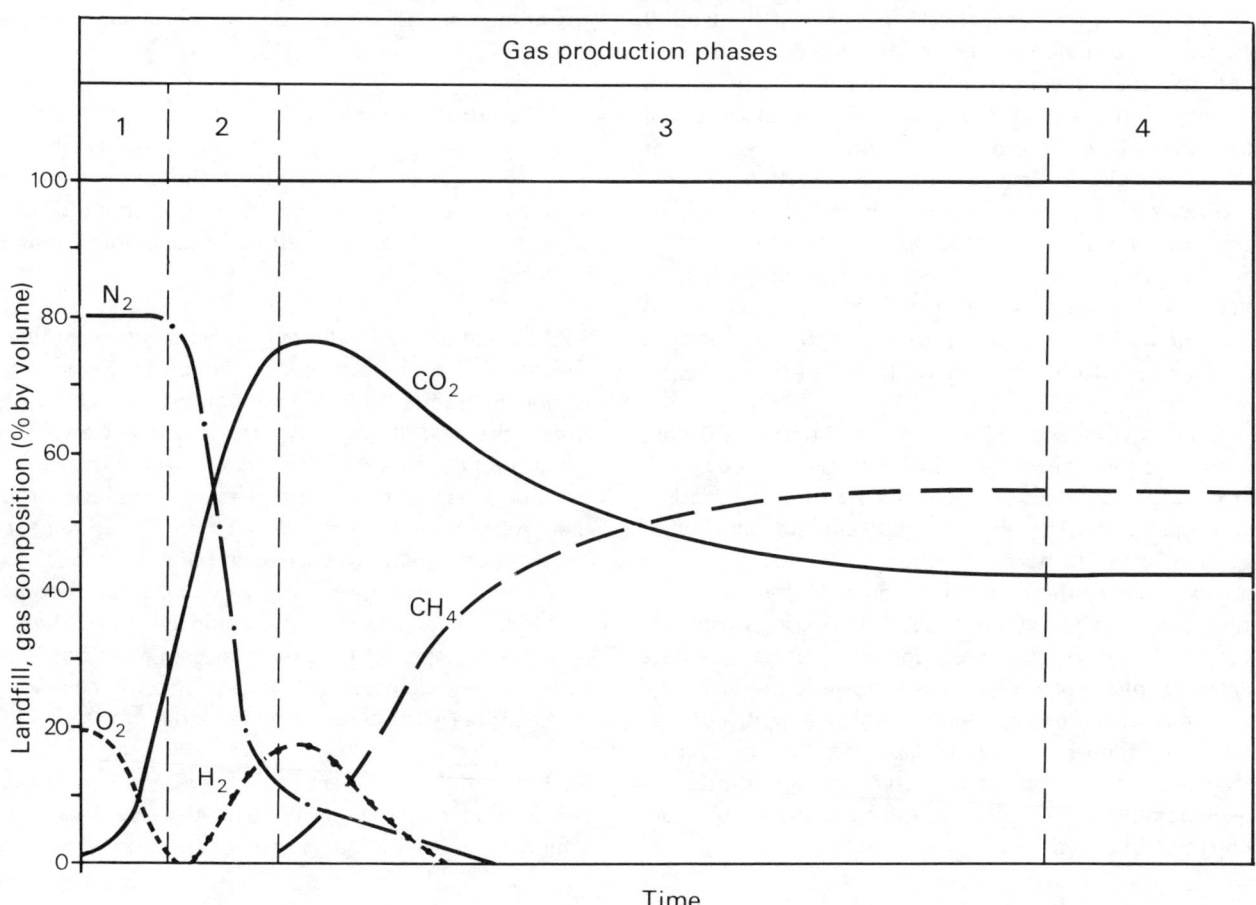

methane production being from pH 6.4 to 7.4. The optimum temperature for anaerobic decomposition lies between 29°C and 37°C. After the waste is deposited the oxygen content of trapped air is quickly utilised. Any further ingress of air, which would inhibit methane generation, is likely to be minimal unless excessive extraction of landfill gas results in air being drawn in. Recirculation of leachate is also unlikely to introduce sufficient oxygen to inhibit gas production. Finally the co-disposal (see paragraph 7.7) of wastes with household wastes may affect the rate of gas production. Some industrial wastes can inhibit gas production by reducing the efficiency of the microbial processes while, on the other hand, co-disposal of wastes with a high putrescible content may enhance gas production.

Composition of landfill gas

3.73 Biological and biochemical decomposition of wastes takes place over a number of years, and during this time considerable changes occur in the nature and quantity of the gas evolved. These changes are described in Annex 2 and summarised in Figure 3.73. Initially, air entrapped in waste when deposited is rapidly replaced by carbon dioxide and hydrogen. The production of hydrogen, which may reach concentrations of approximately 20 per cent by volume of total gas, normally occurs during the first few months. Typically some 6 to 12 months after deposition substantial amounts of methane will start to be produced. The concentration of methane will gradually increase until it reaches about 65 per cent of the landfill gas, with a corresponding decrease in the carbon dioxide content. When the methane has reached its maximum concentration, the methane/carbon dioxide mixture will continue to be evolved over a period of some years. It is difficult to give a timescale for the evolution of these gases or for the period during over which they will be produced. However, at large, deep landfills the production of methane at maximum concentration for periods in excess of 10 years can be expected. At more shallow sites the period will be less and possibly much less if aerobic conditions persist for an extended period of time.

3.74 A large number of minor constituents at low concentrations have been identified in domestic waste landfill gas. Some of these compounds are often responsible for odour associated with the gas, but are unlikely to represent a health hazard. Hydrogen sulphide is unlikely to make any significant contribution to landfill gas odour unless high concentrations of sulphate bearing wastes are present. In the presence of domestic refuse, sulphate sludges or plasterboard have been shown to produce large quantities of hydrogen sulphide, which is both odorous and toxic. Odours from landfill gas differ from those from leachate since the smell of the latter is predominantly due to carboxylic acids which are only present at low concentration in the gas.

Environmental problems arising from landfill gas

3.75 Problems due to landfill gas fall into the following categories:

(a) explosions or fires due to gas collecting in confined spaces, such as buildings, culverts, manholes or ducts on or near landfill sites;

(b) asphyxiation of people entering culverts, trenches or manholes on landfill sites;

(c) when released through fissures at the surface, landfill gas maybe ignited with a risk of setting fire to the waste;

(d) detrimental effects on crops or vegetation on or adjacent to landfill sites;

(e) risks to human health from gas emissions;

(f) nuisance problems, especially odour.

It is apparent from a recent survey that the number of incidents involving landfill gas is increasing. However, to date serious incidents have been very few.

3.76 Means of reducing the potential risks at (a) and (b) above are discussed below (paragraphs 3.80-3.83) and in Annex 2. The risk of causing fires within the waste by igniting escaping gas calls for appropriate care in site operations (see Chapter 5 and in particular paragraphs 5.70-5.74).

3.77 Landfill gas has also been responsible for damage to vegetation particularly on restored landfills. Methane displaces the normal soil atmosphere and prevents diffusion of oxygen from the air into the soil resulting in the soil being rendered anoxic. Phytotoxic compounds may also inhibit plant growth.

3.78 Landfill gas is diluted above a site typically by between a thousand and over a million fold with the air. Although several organic compounds may be present within the landfill gas at significant concentrations these are also diluted to several orders of magnitude below any toxicity threshold but nevertheless may still be odorous as they are released into the atmosphere. Measurements of metals as volatile derivatives or in particulate matter have shown that concentrations are very low at both domestic and industrial waste sites. It should be noted, however, that at sites taking difficult wastes care must be taken to ensure that toxic gases are not generated by the mixing of incompatible wastes (see paragraph 7.44).

3.79 Landfill odours can cause considerable nuisance, as the dilution required to eliminate the smell may not be achieved under some weather conditions. Potential sources of odour include wet refuse which has been allowed to decompose partially before deposition, or large gas vents releasing gas directly to atmosphere. Minimum disturbance and finely graded cover material will do much to alleviate the smell nuisance (see also paragraph 5.68).

Lateral diffusion of landfill gas

3.80 On sites situated in strata of medium to high permeability such as sands, gravels, chalk or fissured rock, gas can migrate laterally for considerable distances beyond the landfill site boundary. While in most circumstances the gas will eventually vent to atmosphere, at some landfills it may collect in buildings situated around the site and create both a potential fire or explosion hazard and environmental nuisance due to its smell. The potential for gas migration should therefore be considered when planning a proposed landfill operation or any development adjacent to a landfill site.

3.81 Three types of system may be used, either singly or in combination, to prevent lateral migration of landfill gas:

(a) impermeable barriers;

(b) passive venting; and

(c) gas pumping.

Materials which have been used to prevent gas migration include puddled clay, bentonite, plastic sheeting and PFA/cement grouts. The placement of such materials and the problems that may be encountered in their use are discussed in detail in Chapter 4. For sites up to 5 metres in depth, a trench excavated to about the depth of the base of the refuse around the perimeter of the site and lined with a barrier material should be sufficient to prevent gas migration. The width of the trench does not appear to be critical. For deeper landfills impermeable barriers should be installed as if they were to prevent leachate migration.

3.82 Alternatively lateral migration of gas is effectively prevented by surrounding the site with a zone of high permeability material through which the gas can readily vent to atmosphere. For sites up to 5 metres deep a perimeter trench filled with coarse aggregate or builders' rubble has proved satisfactory. Care must be taken, however, to prevent sealing of the top surface with mud, snow, ice and the like. Such blockages can be overcome by installing a horizontal perforated pipe in the aggregate filled trench, the pipe being connected to vertical vent pipes at intervals of some 20 to 30 m. With this system the upper surface of the trench may be sealed with clay or similar material. At deeper landfills a passive venting system could be created by depositing coarse aggregate between the side walls and the waste. However, gas release rates may then be high and could cause a nuisance. Alternatively, boreholes with slotted casing extending to

the base of the refuse may be drilled either around the perimeter of the site or within the waste itself. However, unless a sufficient number of such boreholes are installed it is likely that pumping will be required to ensure an adequate rate of ventilation.

3.83 The quality and quantity of intermediate and top cover used in operations and restoration also influences the extent of lateral gas migration. For example the specification of a low permeability cap as part of the restoration may well encourage lateral migration by effectively preventing gas escape.

Landfill gas monitoring

3.84 As indicated above, landfill gas is one of the major routes through which landfilling may impact on the environment and the local community. It is also in some cases a potential source of energy supply. Landfill gas therefore needs to be monitored to ensure that its production does not place a landfill site or its surroundings at risk as a result of gas migration or dangerous concentrations being allowed to accumulate. There is a temptation to combine landfill gas and leachate monitoring at the same borehole. Care must be taken to ensure that any equipment used to monitor leachate is compatible with a landfill gas environment. The practice of lowering pumps into monitoring boreholes to pump out leachate is not to be encouraged at sites where gas is evolving (see Annex 2).

Collection and utilisation of landfill gas

3.85 At some landfills it may be worthwhile extracting gas for use as a fuel but to be successful a number of requirements have to be met. These are:

(a) a suitable use for the gas must be identified,

(b) the landfill must have a minimum depth of at least 10 m of biodegradable material,

(c) there must be a large quantity of waste already deposited. Experience suggests that at least 0.5 million tonnes is required,

(d) the waste should not be too old. Wastes deposited for between 5 and 10 years seem generally to produce the highest gas yields; and

(e) the water level should be at least 5 m below the landfill surface. Saturated conditions are not conducive to landfill gas collection.

By these criteria, venting gas to the atmosphere or flaring will remain the only control option for most landfills. However, at some landfills it may be worthwhile utilising the gas as a process fuel, for electricity generation or, conceivably, as a chemical feedstock.

3.86 A preliminary assessment of quality and potential

quantities of landfill gas is essential and can best be achieved by drilling wells at appropriate locations and determining the quantity of gas that can be extracted at various suction pressures. It is important to determine the sphere of influence for each extraction well and this is usually measured by placing piezometers at various distances from the well. The sphere of influence will vary but usually has been found to be between some 30-50 m radius. The number of wells required can vary, around 2 wells per hectare being an average figure. The rate at which gas can be withdrawn is also variable, a typical figure being about 3 m^3/minute. A typical calorific value for landfill gas is 20 MJ/m^3, corresponding to about 55 per cent methane.

3.87 Usually a landfill gas extraction well has a diameter of between 0.3 to 0.6 metres and is drilled either to the base of the refuse or until the water table is reached. The well casing is usually some 0.1 m in diameter and perforated for the bottom half, the annulus around the lower part being backfilled with permeable material, followed by clay or other sealing material.

3.88 Gas production wells are joined together by plastic pipe, usually of at least 100 mm diameter, which in turn is connected to a suction pump. The pipes, which are often laid on the surface of a landfill, are normally sloped to facilitate the collection and removal of condensate.

General effects of landfill on the environment

Introduction

3.89 Nearly all forms of waste disposal ultimately involve landfill and for most wastes direct transport to landfill sites is the most economic route. The need to ensure that the landfill operation does not adversely affect the environment is therefore of considerable importance. Thus the aim is not to landfill wastes which could cause a long term hazard. It is also important to recognise that landfill operation may affect a wide area and have a substantial impact on local communities. Such impacts will have been considered during the planning application stage; they will have involved consideration of such matters as access, landscape quality, agricultural land quality, restoration and afteruse as well as amenity aspects such as traffic, litter, noise, dust and smell. Local interests should have been carefully studied to minimise adverse effects of the operation on the community. The importance of good relations with the local community cannot be overemphasised.

Landscape

3.90 Landfilling inevitably has a visual impact. Sites may need to be located in scenically attractive areas and in particular a number of existing, and some potential landfill sites are in designated areas such as National Parks or Areas of Outstanding Natural Beauty. In addition, some

landfill sites may contain features of special interest or be sites of special scientific interest (SSSI's) as defined by the Nature Conservancy Council. Particular planning and other policies may apply to such areas or sites. Accordingly, a landfill site should intrude as little as possible on the landscape during its operational life and should not result in the unnecessary loss of features of interest and significance. Before commencing landfill operations, plans should have been approved to avoid adverse visual impact and to ensure that restoration is progressed as quickly as possible. A landfill site may be screened from areas of public access by either a tree or shrub planting scheme or by using natural ground contours to assist in visual screening. A restored landfill should quickly become an integral part of the landscape.

3.91 However, a sensible balance has to be struck. For example if a site contains a geological outcrop of significant interest, it may be possible to preserve it only if the exposure is near the surface of the site or alternatively in an area of the site not to be landfilled. A decision therefore must be made on whether waste disposal or site preservation is to prevail. Such a decision must take account of the location of the site in terms of the waste disposal strategy, the importance of the features it is wished to preserve, and the economic and environmental consequences of alternative options.

Ecology

3.92 Landfill operations usually take place in quarry or mineral excavations, derelict land or less frequently on undisturbed ground or agricultural land. All sites will differ in landscape and ecological values and in other site characteristics but for all of them a landfill operation is likely to have a major, and possibly irreversible, effect on the ecology of the site and its immediate area. A landfill operation will result in the destruction of existing vegetation and a significant disruption to animal life. Only in the most exceptional circumstances will it be possible to design a landfill to preserve, for example, individual examples of trees and shrubs.

3.93 However, such changes must be balanced against both the wider advantages gained from having a landfill site at a particular location and the benefits that will accrue once a site has been restored. Restoration should aim at integrating a landfill site within the existing landscape and establishing a viable ecology, possibly more valuable than the one it replaces. Engineering landfill operations to take account of effects on ecology is a recent development but one which is becoming increasingly important.

Local community

3.94 One of the more obvious effects of landfill on a local community and one which is a continuing source of complaint is the additional traffic generated by its presence. Heavy lorry traffic gives rise to nuisance from noise, vibration, exhaust emissions, dust, dirt and visual intrusion. The size of waste disposal vehicles may also be considered intimidating to pedestrians. Heavy vehicles on narrow roads will lead to delays to other traffic, while damage to road verges and the road surface itself will always be a source of nuisance and complaint.

3.95 Whenever possible heavy vehicles should be routed to reach their destination by travelling on major roads. This can be achieved either by voluntary agreement, or specified in a waste disposal contract where one exists. Use may also be made of relevant provisions in the various Highways Acts. Travelling on minor roads should be kept to an absolute minimum and in some cases this consideration should outweigh that of minimising transport costs. However, because landfill sites are frequently located in mineral workings, some travel on minor roads is likely to be unavoidable though in such cases the volume of traffic carrying wastes may not be too disparate from the traffic formerly servicing the quarry. Some road improvements may however, be needed. These may include strengthening road shoulders, adequate provision of footpaths, the improvement of sight lines and corners and possibly the provision of passing places or laybys. In cases where the landfill has a long life expectancy the provision of bypass roads may be considered.

3.96 It should always be borne in mind that transport by road is not the only transport option. In some cases rail, river, canal and conveyor transport can be viable alternatives.

3.97 Landfill operation can affect the local community by taking place close to housing. During their operational lifetime, landfills, even with the highest standards of management, are likely to be unpopular with people living close to them, even though they may offer positive benefits once restored. It is therefore common practice for a buffer zone to be required between landfills and housing. Alternatively, strategically placed and landscaped embankments may be effective in screening a site from nearby housing. The need for a buffer zone or screens should be a matter for discussion between planning and waste disposal officers, and the local environmental health department: a similar need is likely to arise when housing development is planned near sites that are likely to become landfills. However, buffer zones should not be arbitrarily imposed but should reflect a considered evaluation of the impact of the landfill, account being taken of such things as topography, direction of working, type of plant to be used, method of operation and prevailing wind direction. It is now recognised that it is not sensible to set a fixed width for a buffer zone. In some instances as an alternative it may be possible to arrange for inert waste to be deposited close to housing and thereby create an effective buffer zone without forfeit of waste disposal opportunity. Sites situated close to housing should, as far as possible, be completed quickly. Progressive restoration should be an integral part of the operational plan.

3.98 Landfill operations are a source of concern to the local community through potential nuisances such as litter, vermin, flies, dust, odours, fires and noise. Each of these can, however, be controlled and their effects minimised by good management and there is no excuse for them having a significant impact on the community. Methods of dealing with these problems are discussed in detail in Chapter 5.

3.99 Reservations the public may have about landfill operations tend to fall into three categories:

(a) reduction in property values in particular and impact on the community in general;

(b) impacts on their health and on the environment;

(c) the competence and financial probity of the operator.

Such reservations may or may not be legitimate but cannot be dismissed out of hand. Any potential operator should take every opportunity to show why the landfill is necessary, what the alternatives are and their costs. He should discuss the likely impacts of the proposed operation and the steps he proposes to take to minimise them and the benefits the restored site may represent to the community in the long term. The operator should also be in a position to demonstrate, via slides, films and models, the progressive stages of development the landfill will undergo. Where possible he should also give examples of other operations he has conducted. Above all, the operator should be sympathetic to community reaction and be seen to act, whenever possible, in reducing public concern.

3.100 All planning applications for landfill need to be advertised and this gives the local community an opportunity to express its views. Advertisement is not however, necessary where landfill is to be undertaken as a permitted development. The disposal licence application need not be advertised, but a pre-requisite of issuing a licence is a valid planning permission. The way in which the local community first hears that a landfill operation is planned or is about to start can set the tone of the relationship between the community and operator. While it may be the policy of the local authority to inform residents close to the site about proposed developments and to meet with Residents Associations and Community Councils, the site operator should also be playing an active role. It will be to his advantage in the long-term to establish a dialogue with the local community by informing them of his proposals in advance, when he has applied for a disposal licence and when he is about to commence work. This will be of particular importance at sites which have longstanding planning permission. The need for good dialogue applies equally if the site is to be operated by the local authority or a private contractor.

3.101 Often communities have Residents' Associations or Community Councils and the operator should liaise with them to keep himself informed of local opinions and to keep the community appraised of developments at the site. If the site's location may give cause for concern a group comprising the operator, local authority, and community representatives should meet on a regular basis to prevent issues from developing into conflicts. On occasions local groups have suggested beneficial improvements to site operations and restoration schemes.

3.102 In endeavouring to achieve a good relationship with the local community and demonstrate his competence, the operator should give prompt and sympathetic attention to complaints. By dealing with them quickly the operator will demonstrate that he is determined to operate his site efficiently and is taking due care to safeguard the environment and the community.

Site investigation

Introduction

3.103 Good landfill management starts with a site investigation. The objective of the investigation is to produce accurate information on all ground conditions in the vicinity of a site which may influence effects on the environment. Such information is a pre-requisite to the design, development, monitoring and restoration of a landfill. Site investigation should proceed in stages. It starts with a reconnaissance study the aim of which is to establish the need for, and feasibility of, landfill operations in the area, and to identify possible sites. An essential consideration at this stage is the level of investment likely to be needed and the viability of the projected operation. If the reconnaissance study is promising there will follow a detailed field investigation stage. The report of this investigation will provide the basis for the landfill design and for the formulation of the working plan for the site.

The economics of landfill waste disposal

3.104 Waste disposal to landfill is a complex and inherently long-term operation, with corresponding difficulties for the prospective operator in fully assessing the financial implications of a proposed development before reaching a decision on whether to proceed. However, it is clearly not in the interests of either the prospective operator or the environment (including the local community) that a landfill development should proceed on a shaky financial basis. There is at stake, in the long run, not only the success of individual developments but public acceptability of landfill as the principal disposal option. It is therefore of considerable importance that the economic aspects of landfill operations should be clearly understood and that in reaching decisions on the investment required, prospective operators should make proper provision for efficient management, with due regard to environmental concerns and the needs of site restoration and aftercare.

3.105 The economic assessment of alternative sites or disposal options calls for the application of appropriate costing techniques which enable proper account to be taken of the likely incidence of costs during the long life of a landfill. A detailed description of such techniques and their application is given in Annex 4.

Reconnaissance study

3.106 Initially the potential operator must establish the need for a landfill and evaluate the types of waste being generated in the area. If such a need is identified and the waste arisings are of types which the operator wishes to handle, the level of investment needed to dispose of wastes in an environmentally acceptable manner requires consideration. The importance of this assessment is described above. The potential operator will need to study the county structure and waste disposal plans for the area together with any local plans (see paragraph 2.5 et seq), and should then engage in preliminary discussions with the planning, disposal and water authorities to obtain their views on planning control, waste disposal and water pollution aspects of the proposition.

3.107 The reconnaissance study involves both a desk study and field inspection; its aim is to identify a few sites which may be suitable for the proposed waste disposal operation and which appear to merit a more detailed site investigation. The study should deal with the location of the proposed operation in the overall planning stragety of the county and in the context of, land use policies, the likely environmental and social effects of the proposed operation and the potential impact on water resources. Full use should be made of available data; examples of the type of data that might be used and the relevant sources are shown in Table 3.107.

3.108 The desk study must be supported by a visual inspection of the sites. During this inspection, particular attention should be paid to three particular aspects. First note should be taken of the environment surrounding the site, the land use practices and size and proximity of communities. Second the existing transport network should be evaluated in the light of the vehicle movements expected to and from the site. Finally attention should be paid to the potential of the site to affect water quality adversely. Particular attention should be given to any exposed strata at or near the site and the existence of watercourses, springs and areas of standing water. A rough sketch map of the site and its surroundings should be made to show such features. If the site was originally a quarry, then details of stockpiled material, plant and equipment and land use on and around the site should be included.

Detailed site investigation

3.109 Detailed site investigation proposals should be discussed with the relevant authorities. These discussions should aim to identify the type of information they will

Table 3.107 Information sources for preliminary reconnaissance

Subject	Source	Information
Topography	Ordnance Survey topographic maps Development plans Quarry plans	Relief (approximate ground levels) surface water drainage. Proximity of housing. Access.
Geology	British Geological Survey Maps Sheet memoirs British regional geology Field records and borehole records Quarry records Mining records	Geological succession (bedrock and superficial deposits) thickness of strata and lateral extent. Geological structures — attitude of strata and major discontinuities (eg faults). Mineral resources and utilisation.
Hydrogeology and Water Supply	Regional Water Authority Site Operators	Extent of water utilisation (river and reservoir catchment areas and aquifers). Distribution of licensed water abstractions. Significance of on-site surface groundwater.
Climate	Meteorological Office	Long term average rainfall and potential evaporation for calculation of effective rainfall and leachate generation.
Agriculture and other existing land uses	Soil Survey c England and Wales MAFF (ADAS) DAFS Nature Conservancy Council	Soil series and characteristics; agricultural land classification. Other land uses and ecology. Nature conservation.
Landscape	Countryside Commission	Countryside policies; particularly National Parks and Areas of Outstanding Natural Beauty.
Planning Policies	Structure and Local Plans	Areas of where landfill operations may be acceptable.

require in support of planning and disposal license applications. An assessment of the impacts of the operation on the environment, local communities and the transportation system is also likely to be required. It should be noted that certain aspects of site investigations, such as drilling boreholes, may require planning permission.

Hydrogeological investigations

3.110 An adequate appraisal of the geological and hydrogeological conditions at the site will be necessary. This should aim to provide details of soil permeabilities, groundwater levels and variations in the water table, depth of unsaturated zone, mineralogy of the strata and other information necessary to assess both the probable rates of leachate movement and the likely attenuation of pollutants that will occur.

3.11 All investigations should be supervised by a competent hydrogeologist. If necessary a detailed contract specification and schedule of work should be provided. The specification should indicate whenever possible the

conditions expected at the site, the number of boreholes required and the proposed depth of penetration. It should however be flexible enough to make allowance for hydrogeological features that become apparent as the work proceeds. For this reason the specification should give sufficient information for contractors to submit a realistic schedule of rates. A bill of quantities should not be used, or used with caution.

3.112 The basic specification for a hydrogeological investigation should include the following:

(a) drilling method;

(b) sampling and field testing requirements;

(c) laboratory testing requirements;

(d) methods to be used and materials for borehole completion.

A clear distinction should be made between those items of

sampling and testing which can be carried out by the contractor and those to be undertaken by the supervising hydrogeologist.

Borehole drilling and completion

3.113 BSI Code of Practice for site investigations (BS5930-1981) should be followed as far as possible. Where unconsolidated materials are to be penetrated the use of percussion drilling methods is preferred. Temporary steel casing should be installed progressively and the borehold advanced without the addition of water, since important hydrogeological features may otherwise go unrecognised. Each water strike should be accurately recorded and adequate time allowed (say 15 minutes) for rest water level to be achieved and measured. Temporary casing should be kept as close as possible to the base of the borehole to minimise vertical mixing of groundwater.

3.114 Where consolidated strata are to be penetrated allowance should be made for the use of rotary drilling. The use of air as the flushing medium should be specified wherever possible, although it should be noted that this technique can cause short term dust problems. Where rapid progress is desirable, open hole drilling with a rock bit may be considered expedient. In this case the drill cuttings should be examined continuously to determine stratigraphic sequences. Geophysical logging tools may be needed to provide supplementary information. Rock cores should be extracted wherever necessary to identify the rock type and to obtain samples. Sampling is usually conducted at 1.5 m intervals and when any water inflow occurs, and should be undertaken by a qualified hydrogeologist.

3.115 Once boreholes have been advanced to the required depth they should be completed with the installation of one or more piezometers. Borehole completion should be properly designed and take into account the information gained during drilling. Wherever boreholes are to be retained for future groundwater sampling, standpipes of at least 50 mm diameter should be provided to permit access. Where boreholes are to be used solely for water level measurements, 19 mm standpipes may be sufficient. In situations where it is deemed necessary to monitor groundwater at depth, the use of a slotted standpipe surrounded by an annular gravel pack may need to be provided.

Other site measurements

3.116 During site investigations it may also be necessary to provide information on other features such as:

(a) detailed mapping and sampling of exposed soils and rocks;

(b) mapping of all surface streams, springs and areas of standing water;

(c) flow rate measurements along surface streams to determine zones of stream loss or groundwater discharge. (Such information may be obtainable from the Regional Water Authority or River Purification Board.); and

(d) evaluation of the type and quantity of materials for intermediate and final cover and of restoration materials available on site.

Additionally, at sites in areas with an extensive fluvio-glacial overburden, it may be necessary to obtain more detailed lithological information. In such cases the use of surface geophysical techniques or shallow power auger borings may be required.

Surveying

3.117 An accurate ground level survey will be needed for landfill site design and to calculate the voidspace available. Additionally, surveying will be required to provide datum points for each borehole and other site features such as stream beds, springs, outcrops and exposures.

Water level measurements

3.118 Water levels should be recorded at least daily for the duration of the investigation. Thereafter, water level measurements should be made every week for four weeks and thereafter at monthly intervals. Relevant daily climatological data should also be obtained from the nearest weather station throughout the investigation period.

Site investigation report

3.119 The report of the site investigation will provide information that is vital to the design of the site so as to ensure the protection of water quality, and also provide essential information necessary for formulating the working plan (see below). The knowledge of site characteristics obtained from the investigation enable basic features of site development and operaton to be specified, including:

(a) site preparation work;

(b) water pollution prevention measures;

(c) pollution monitoring;

(d) gas venting needs;

(e) intermediate cover requirements;

(f) quantities of top soil, sub-soil and soil making material available on site;

(g) estimated quantities of soil needed for restoration and,

(h)   vehicle and machinery requirements.

The report thus provides a clear indication of the resources the operator will need to allocate to develop and run the site.

The working plan

3.120   The working plan is the basic document needed to support disposal licence and planning applications, as well as for operation of the site. The plan should consist of two parts:

(a)   a plan or series of plans, outlining the development, and

(b)   a description of the way operations are to be carried out.

Plans

The plans required are as follows:

(a)   *Site location plan* — on a suitable scale to identify the site in comparison with the Ordnance Survey Map of the area.

(b)   *Site survey plan* — showing the landfill site prior to preparation including existing ground levels or contours, existing drainage and outfalls, the location of any watercourses, unstable ground and slips, adits, shafts and other features of the site.

(c)   *Ground investigation summary* — giving geological survey information such as borehole data, trial pits and other ground investigations.

(d)   *Site operational plan* — showing the site after its preparation including:

(i)   gates and boundary fencing,

(ii)   access and position of notice board,

(iii)   reception facilities, site control office and other buildings or fixed equipment on site including wheelwasher and weighbridge,

(iv)   location of site compound and provision for parking,

(v)   location of storage tanks,

(vi)   vehicle circulation routes,

(vii)   phases and direction of filling,

(viii)   emergency or seasonal tipping areas,

(ix)   location of cover material,

(x)   location of bunds and lining,

(xi)   new drainage layout,

(xii)   leachate treatment disposal,

(xiii)   gas control facilities,

(xiv)   the location of facilities for the storage and disposal of difficult wastes and

(xv)   landscaping and screening works.

(e)   *Engineering plans* — showing details of any engineered features. The plans need not be to the same standard of detail as that required for construction but should show clearly the intended size, level and materials of construction for the proposed works. Examples of works which may require these details are:

(i)   the form and construction of bunds,

(ii)   details of surface water drainage,

(iii)   details of groundwater engineering,

(iv)   form and construction of site roads,

(v)   any access improvements,

(vi)   any site maintenance facilities,

(vii)   wheelwash and weighbridge facilities,

(viii)   site control office and mess facilities,

(ix)   leachate collection, treatment and disposal facilities,

(x)   facilities to deal with and dispose of difficult wastes, and

(xi)   gas control measures.

(f)   *Restoration plan* — showing the form and contours of the restored site, including:

(i)   phasing of restoration,

(ii)   final contours after settlement,

(iii)   final drainage patterns,

(iv)   cross-section(s) through the restored site,

(v)   details of the final cover or capping of the site and its possible source and

(vi)   details of proposed afteruse and planting.

Descriptive material

3.122   Each drawing should include annotations and side notes. The drawings should be commensurate with the type and scale of the facility and should be explained by an accompanying statement of intent which should include the following information:

(a)   an outline of the operational principles of the landfill,

(b)   an outline of the site preparatory works to be carried out,

(c)   procedure to be followed to record waste input at the site,

(d)   the method of operation to be adopted on site,

(e)   environmental control measures to be adopted,

(f)   water pollution control measures,

(g)   environmental monitoring arrangements,

(h)   procedures to check the authenticity of delivered wastes,

(i)   procedures to be followed to deal with unacceptable wastes delivered to the site,

(j)   an outline of the final restoration and aftercare of the landfill,

(k)   any provision being made for gas control or extraction during and after filling and

(l)   an inventory of equipment to be used on site.

MICROBIAL DECOMPOSITION (see paragraphs 3.22 and 3.73)

Biodegradation of components in refuse (see Figure 3.73)

A.1 Aerobic conditions will prevail initially. However, anaerobic conditions are rapidly established with methane and carbon dioxide being the final products of carbon metabolism. Carboxylic acids are major intermediates and can remain as the end products of carbon metabolism. The metabolism of landfilled domestic refuse is dominated by carbohydrates; these constitute about 50% of the dry weight; proteins and lipids account for only 5 to 10% of the dry weight of refuse in the UK.

Decomposition of carbohydrates

A.2 Carbohydrates decompose initially into simple sugars such as glucose and cellulose which can then be further degraded to form hydrogen and carbon dioxide as well as acetic, propionic, butyric, valeric and caproic acids. These carboxylic acids are major pollutants in leachate from waste which has been deposited for less than 10 years. Of the carboxylic acids acetic acid is a major constituent. For example, leachate from freshly deposited waste contained the following concentrations of acids: acetic 3,800 mg/l; propionic 1,600 mg/l; butyric 3,500 mg/l; valeric 2,100 mg/l and caproic 3,700 mg/l.

A.3 The cellulose found in domestic waste consists of materials of differing biodegradability. Thus cellulose present in landfilled paper will degrade more quickly than that occurring in garden waste. However, it should be emphasised that, particularly if the water content of waste is low, cellulose may take many years to degrade.

Protein decomposition

A.4 It is probable that the metabolism of proteins in landfill follows the same course as in anaerobic sewage sludge digestion where hydrolysis to peptides and amino acids occurs. Amino acid deamination leads to the formaion of short chain carboxylic acids, carbon dioxide and ammonia. Amino acid deamination is the sole source of the isobutyric and isovaleric acids found in leachate.

A.5 Protein decomposition is the major source of ammonia found in landfill leachate. Concentrations of ammonia in leachate from freshly deposited waste typically range from 500-2,500 mg/l, while concentrations in liquid associated with older waste are lower ranging from 100-500 mg/l.

Methanogenesis and sulphate reduction

A.6 Hydrogen is normally found only for a few weeks after refuse deposition but may be produced for longer periods when the growth of microbes is inhibited by dry conditions. The presence of hydrogen is also prolonged in wet pulverised refuse where rapid production of high concentrations of carboxylic acids (about 40,000 mg/l total acids) inhibits methanogensis and sulphate reduction. Sulphate concentrations in leachate are often between 2,000 and 3,000 mg/l, but evidence of enhanced rates of sulphate reduction has been found at some landfills with residual concentrations as low as 50 mg/l. At such landfills low sulphate concentrations are often accompanied by low concentrations of carboxylic acids and with correspondingly high rates of methane production.

Landfill microbiology

A.7 Most studies of landfill microbes are dominated by public health considerations. Attention has been focussed on the presence of organisms of faecal origin and viruses in refuse and leachate. Only limited data exists on the number or physiological activities of organisms involved in the decomposition of domestic refuse in landfill.

Temperature effects

A.8 Landfill temperature is determined by the rate of energy production within the landfill and the rate of heat loss to the environment. The optimum temperature for refuse decomposition and methane production has been shown to be about 40°C. This temperature is not normally attained in shallow landfills where temperature fluctuations follow ambient levels. During refuse placement, the temperature can rise to about 30°C initially as a result of aerobic metabolism. In deeper landfills, the temperature can rise to 60°C and temperatures of 40-45°C can be maintained over the first five years of the operational life of the site.

Water effects

A.9 The role of water in the decomposition of waste is important. In general, as the percentage of water increases so does the rate of decomposition. The rate of gas production at moisture contents of 30-40 per cent wet weight generally is less than 10 l/tonne of refuse per day. At higher moisture contents the rate of gas production increases rapidly. Optimum methane production occurs at 50 to 60 per cent water content.

Waste density effects

A.10 In-place density affects the decomposition processes. If waste is deposited without compaction (0.35 tonne/m³), air ingress into the mass can readily occur. Decomposition is then essentially aerobic, the principal products being carbon dioxide and water with temperatures greater than of 80°C being observed. At refuse densities in excess of 0.5 tonne/m^3 the dominant decomposition process becomes anaerobic with the appearance of carboxylic acids in leachate and methane in the gas phase.

A.11 As the in-place density of waste is increased, water penetration will be more difficult and the overall moisture content will not rise significantly. Since the rate of decomposition increases with increasing moisture content, high compaction will tend to lower the rate of decomposition. However, in some situations compaction to a high density may mean that saturation occurs with lower liquid inputs. High levels of gas generation have been observed at in-place densities of 0.8 t/m^3 or more.

CHAPTER 4: LANDFILL SITE DEVELOPMENT

Introduction

4.1 Development of a landfill may begin only after planning permission has been obtained and the disposal licence has been issued. A working plan for the landfill (see paragraph 3.120) will have been agreed between the operator and the licensing authority and the disposal licence will normally stipulate that, prior to any waste deposits, the designs, works and procedures specified in the working plan must be implemented. The preparations needed at a site before waste can be deposited may be considered under two headings: the infrastructure, that is, the position of the building, roads and facilities that are necessary to the efficient running of the site; the basic engineering works needed to shape the site for the reception of wastes and, generally, to meet the technical requirements of the working plan.

Site infrastructure

4.2 The infrastructure provided at a landfill sets the general framework within which the site will be operated. Its careful planning is essential if the landfill is to operate efficiently and safely throughout its operational phase.

Site entrance

4.3 At many sites, and particularly at operational or recently abandoned mineral workings, access arrangements will already be established (although it should be noted that during landfilling operations the volume of traffic using the access may be greater than it was during mineral extraction). At other sites access will have to be provided. In either case an adequately splayed bellmouth, properly surfaced and capable of taking traffic in both directions will be required. Road safety considerations may necessitate the provision of an entrance set back from the highway with good sight lines and possibly an acceleration and deceleration lane. Vehicles using the landfill should not be required to queue on the highway. The design of the site entrance will depend on a number of factors including the number and type of vehicles using the site and the classification of the highway from which access is to be gained. Landfill operators should consult with the local authority's traffic engineers before making their planning application.

4.4 The appearance of the entrance is a major influence on how a landfill site is perceived by the public. An untidy entrance with mud on the roads and highway does little to instill confidence in the operation being undertaken. Judicious landscaping and tree planting at the site entrance will provide a screen for the operation and be a pleasing feature.

Site accommodation

4.5 The size, type and number of buildings required at a landfill depends on factors such as the level of waste input, the expected life of the site, environmental factors and the availability of other facilities and depots. Buildings will therefore range from single portable cabins/messroom units to purpose built complexes incorporating offices, messrooms, stores, garages and workshops. However, regardless of the number of units to be provided, certain features are common and should be borne in mind. These include:

(a)   the need to comply with planning, building, fire, and health and safety regulations and controls,

(b)   security and resistance to vandalism,

(c)   durability in service and the possible need to relocate accommodation during the lifetime of the site,

(d)   ease of cleaning and maintenance,

(e)   appearance, and

(f)   the availability of services such as electricity, water, drainage and telephone.

4.6 Accommodation provided at a landfill should meet the needs of site control and record keeping, welfare and possibly stores, garages and workshops. At large landfills and particularly those taking difficult industrial wastes, provision may also be needed to accommodate management, technical support staff and a small laboratory for check analysis of waste on its reception.

Site control office

4.7 All landfill operators need to control and keep records of vehicles entering and leaving a landfill site. Generally a site control office is needed to achieve this. The type, size and location of the office will depend on factors such as:

(a)   the use made of the site by vehicles not operated by the landfill operator,

(b)   whether the site is within a factory curtilage,

(c)   whether a weighbridge is installed,

(d)   whether other uses are to be made of the accommodation.

In most cases the control office should be located away from the entrance to allow traffic to queue off the highway. At busy sites, and particularly those with a weighbridge it is beneficial to position the office on a kerbed island, ideally to the offside of approaching vehicles. In designing the control office, consideration should be given to providing the control point with a barrier or traffic light system to regulate entry into the landfill. At small sites a combined site control office and accommodation unit will usually suffice.

Welfare/messroom unit

4.8 Except when welfare and messing facilities can be provided close to the landfill such as at a works canteen, special provisions for the welfare of site employees should be made. Facilities may be incorporated as a separate room or areas within the site control office, but at large sites they are best provided in a separate unit. The following facilities should be provided:

(a) lockers for clean and dirty clothing and other belongings together with drying arrangements and separate storage for safety equipment,

(b) adequately heated and lighted accommodation containing a table and chairs or benches for use during meal breaks,

(c) facilities for heating food and providing hot water,

(d) adequate first aid equipment,

(e) a wash basin with hot and cold water,

(f) shower facilities,

(g) sufficient lavatories both for employees and visitors, and

(h) accommodation for the crew of household waste collection/delivery vehicles at sites where site rules allow only the driver to proceed to the disposal area.

Stores

4.9 Space should be provided for the storage of materials used on-site. Potentially harmful substances such as insecticides and weedkillers, flammable substances and liquified gas containers will require special facilities to be provided. Other materials such as diesel fuel, oils and greases should be stored in correctly marked tanks or containers. The Fire Prevention Officer should be consulted over the provision of storage facilities at landfill sites.

Garages

4.10 At some landfills garage and workshop facilities for the equipment may be provided. If a garage is to be used for maintenance purposes it will require adequate lighting, heating, ventilation and insulation. Consideration should

be given to providing a low voltage electricity supply for hand tool operation (see also paragraph 2.77).

Plant and vehicle cleaning

4.11 Arrangements should be made for plant to be regularly cleaned. Areas set aside for this require a good water supply, adequate drainage and specially protected electrical equipment.

Additional accommodation

4.12 At large landfill sites, accommodation should be provided for management meetings and for a display of site development, operation and restoration plans.

Recording waste input

4.13 Reliable site monitoring is dependent on an accurate record of inputs. A weighbridge at the site entrance is the best way of providing such data. Alternatively void-usage may be surveyed on a regular basis.

Weighbridges

4.14 In selecting and siting a weighbridge there are a number of considerations to be borne in mind. Unless the landfill is to be restricted to regular users, vehicles will have to be weighed both going into and out of the site, and this might cause congestion. The weighbridge must also be located far enough away from the site entrance to prevent queueing on the public highway.

4.15 The type of vehicles using the site will determine platform size and capacity. Two types of platform weighbridge are in general use. One type is flush with the road surface; the installation of which calls for fairly extensive civil engineering work. The other type is mounted on the road surface and requires much less engineering works. However, because the bridge can be some 350 mm above the level of the road, ramps must be provided. This type of unit is moveable and can be installed at other sites at reasonable cost.

4.16 Any weighbridge installed at a landfill site is subject to inspection by a local authority trading standards officer and will require to meet the prescribed accuracy standard. Where accuracy is not of paramount concern, the installation of axle weighers may provide a relatively low cost method of weighing. The weighers are usually installed in a shallow pit. Vehicles can be weighed without stopping as the weighers can be linked to a remotely sited load indicator.

4.17 Read-out of weight can be in the form of a digital display or printed ticket. The introduction of computer based technology provides the opportunity to link the weighbridge with a data processing system adapted to the individual needs of the operator.

Wheel-cleaning

4.18 Landfill sites operate in all weathers. Given the nature of the operation, vehicles could carry mud and debris onto the highway. This is a breach of the Highways Act, is unsightly and likely to cause adverse public reaction.

4.19 One means of ensuring that mud is not taken onto the highway is to provide site roads constructed to both a high standard and of sufficient length that any mud and debris trapped on vehicles is then likely to be shaken onto these roads where it can be regularly swept up. However, in many cases site limitations will require the provision of mechanical wheel cleaning equipment. Nevertheless, it is still essential that any wheel cleaning equipment is positioned far enough into the site to leave an adequate length of access road between it and the site entrance to enable any retained mud or slurry to be removed before vehicles drive onto the highway.

Wet spinner

4.20 This equipment comprises either single or twin rollers set in a steel frame mounted over a pit. It may be either sunk to the level of the road or be raised with ramps at either end. A water supply is required. The spinner is operated by the vehicle being driven onto the rollers. This releases the roller brakes, transmitting a driving speed of some 30 mph to the wheels sitting on the rollers, and also opens the water supply valve. The rotation of the rollers draws clean water through pumps which is sprayed onto the wheels and underside of the vehicle.

4.21 A major disadvantage of this system is that it is prone to freezing during winter months. It can also easily be by-passed or misused. Attention must also be paid to the draining of the mud slurry generated at the spinner. A free draining system and overflow incorporated within the design of the pit is preferable, though often recourse may have to be made to pumping the slurry away and periodic digging out of the accumulated mud. Any drainage system installed must be carefully designed so as not to cause pollution of water or freeze in winter.

Dry spinner

4.22 A less expensive method of wheel cleaning is a dry spinner. Removal of mud from the vehicle driving wheels only is achieved by a combination of centrifugal force and the jarring effect of the rollers. The effectiveness of this technique and installation of this device is therefore recommended only when at least 100 m of surfaced access road is also available after the spinner.

Sunken bath

4.23 This relatively low cost system of wheel cleaning involves the provision of a shallow trough, formed in concrete or a similar material, filled with water and through which all traffic leaving the site must be driven. A series of humps or bars set into the road surface within the bath provides vibration which assists in shaking off loose material. Care must be taken to ensure that vehicles do not enter the bath at too high a speed.

Elevated shaker bars

4.24 These are essentially an oversized cattle grid set level with the road, or alternatively elevated with a ramp at either end. The vibration from vehicles passing over the bars causes the mud and trapped material to drop through the grid bars into a trough or sump. The sump requires cleaning at regular intervals.

Underbody spray equipment

4.25 A more expensive system of vehicle cleaning involves the installations of side and underbody water washing spray equipment. This comprises a steel frame mounted over a pit on which pumps, tanks and control equipment are installed. Vehicles are driven slowly through the machine which automatically starts high pressure spray jets aimed at all underbody surfaces, mudguards, wheels and side members.

Safety

4.26 It should be noted that serious accidents involving the use of wheel cleaning equipment have occurred. It is therefore essential that a safe system of work and adequate training and supervision is provided (the Health and Safety Executive can provide detailed advice).

Site notice boards

4.27 It is important that adequate information on the operation and regulation of a landfill is displayed at the site. At the site entrance a notice board of adequate size should be provided, located in a position where it can be read, to give site users and other interested parties the following information:

(a) the site name,

(b) the site operator's name, address and telephone number,

(c) the operator's local agent's telephone number for emergency use,

(d) the name of the Waste Disposal Authority responsible for the disposal licence, and

(e) the opening hours of the site.

4.28 It is essential that users and visitors to the site are fully informed of the site health and safety procedures. In general, agreement on these procedures should be reached by discussions between the site operator and contractors using the site. The key points of the agreed procedures should be clearly displayed on a notice board in a position where it will be seen by all people entering the site. Suitable speed restriction, direction and other signs such as "No Smoking" should be positioned as necessary throughout the site. Signs to control and regulate traffic should comply with the Road Traffic Regulation Act 1967 and other safety signs should comply with BS 5378 (see paragraph 2.75).

Disposal of bulky household waste delivered by the public

4.29 Though not strictly falling within the subject of this review, it is recognised that it may sometimes be desirable, particularly in rural areas, to provide a disposal facility for direct delivery by the public of bulky items of household waste near the entrance to the landfill (such facilities were formerly known as Civic Amenity Sites). Such a facility should be in a separately fenced area situated near to the entrance of the site. It may be provided in association with the local disposal authority to satisfy their obligation under the Refuse Disposal (Amenity) Act 1978.

Roads

4.30 Good quality roads are an essential requirement for efficient landfill disposal operations. Access roads to both the site and disposal area must be maintained in good repair at all times. A suitable traffic flow system should be devised at the planning stage and at busy sites, should preferably include a one-way system around the site.

4.31 The access road from the highway to the site control office is likely to remain in use throughout the life of the site and accordingly should be designed and constructed to meet such requirements. It should be of sufficient length and width to allow vehicles to both pass and queue on it and be surfaced with either tarmacadam or concrete. Suitable road markings and signposts should be used (see paragraph 4.28 above).

4.32 The secondary site roads between the site control office and landfill area will probably require to be relocated during operations as new areas are brought into use. Thought needs to be given at the preparation stage to the location of such roads to ensure that maximum utility and life is obtained. Secondary roads should be wide enough for two vehicles to pass or, where only a small number of vehicle deliveries are expected at any one time suitable passing bays should be provided. Care should be taken to ensure that roads are adequately drained. Operators of busy sites with a large number of vehicle movements should consider providing secondary roads of a high standard. At some landfill sites temporary roads may be required to allow access by vehicles to the operational area. Several designs of moveable temporary roadways are available (see also paragraph 5.12).

Security fencing

4.33 Except where a landfill site has a distinctive and effective natural barrier, adequate fencing should be provided to prevent unauthorised access to landfill sites. Unrestricted access presents significant health and safety problems not only for trespassers but also to authorised site users and occupants of adjacent properties. Generally speaking all landfills should be fenced along their boundaries regardless of the range of waste received. Other areas of a landfill may also require to be fenced and appropriate warning notices posted, for example, where valuable equipment is stored or where liquid or difficult wastes are being deposited. Fencing requirements can thus range from simple demarcation tapes to full scale security specifications. Secure gates are as important as fencing and should therefore match the fencing specification used. Both fencing and gates should be maintained in good repair at all times.

Litter fencing

4.34 Litter control is essential. If permanently fixed litter screens are to be employed preparatory work will be necessary. Developments in recent years have improved the effectiveness of both high and low litter fences. High litter fences, are usually erected around a large area of the site where several months of landfilling may take place in an exposed location. Low litter fences are erected adjacent to landfilling operations and are moved or extended on a day to day basis. The use of both high and low litter fences in combination has proved particularly effective at a number of sites. For operational aspects of litter control see paragraph 5.57 et seq.

Statutory undertaker's equipment

4.35 During the planning of a landfill site, consideration will have been given to the possible effects of the development on equipment (for example, pipeline or cables) owned and operated privately or by public utilities on or adjacent to the site. It is necessary, at an early stage in site preparation, to contact those statutory undertakers whose equipment will be affected by the operation. A list of statutory undertakers who may have to be consulted is given in Table 4.35. Site developers should also note that privately owned pipelines and cable may be present within the site. In such cases local investigations may be necessary to provide detailed information.

Quarry working

4.36 Many landfill sites are in active or abandoned quarries. It is essential that quarry sides remain safe during landfilling activities. Strict controls over the conduct of mineral excavation have been introduced in recent years to prevent quarry sides being left in an unsafe condition, but such controls may not apply to quarries which have

Table 4.35 Statutory undertakers to be consulted before commencing preparation works

Local Electricity Board	Identification of underground and overhead cables, guidance on statutory safe working distances from overhead power lines and protection of pylons and poles.
Central Electricity Generating Board (CEGB)	Diversion of major overhead power lines and pylons would be a major project undertaken only by the CEGB. Consequently adequate funds and timescales should be built into the programme for site preparation should such action be required.
Regional Gas Board	Identification of underground gas pipelines.
British Gas	British Gas are responsible for the provision of the national grid of major gas pipelines. Proposals for landfilling should be examined in regard to existing pipelines and any proposed in the foreseeable future.
British Telecom	Identification and possible diversion of underground cables, overhead wires, telegraph poles etc.
National Coal Board (NCB)	In areas of existing and previous coal mining the NCB should be consulted in regard to the existence of capped and uncapped mine-shafts. In the latter case, site preparation works may involve a capping scheme.
British Airports Authority	Preparatory work in areas near to aerodromes shown on the safeguarding map where for example high jibcranes or tall drilling rigs are to be used.
Water Authority (RWA)	The water authority is a statutory consultee in respect of the effect of the proposals on possible pollution of surface and ground waters including land drainage. The water authority should also be contacted to identify any pipelines (water mains, aqueducts, sewers, culverts and the like) which exist within or adjacent to the site.
Local Authority (Engineers Department)	The local authority will be responsible for street lighting, and may have cables in the vicinity of the site.
Inland Waterways Authority	Landfilling adjacent to a canal will require consideration of drainage patterns and ground and embankment stability.
British Rail	Landfilling adjacent to a railway will require consideration of drainage patterns and also ground or embankment stability.

See also paragraph 2.14 et seq.

been abandoned for some time. Operators planning to infill quarry workings should seek specialist advice on the stability of the sides, overburden and spoil heaps, adits and shafts and on the appropriate measures to be taken. The Mines and Quarries Inspectorate of the Health and Safety Executive should in any case be contacted. They may also be able to indicate where specialist advice may be obtained.

4.37 Periodic inspection of quarry sides during the operation of a landfill is advised. It must be noted that any movement of spoil or work on the faces of a quarry is a quarrying operation and a qualified quarry manager will therefore need to be appointed.

Site engineering

Earthworks

4.38 The working plan may require extensive earthworks to be carried out before deposition of waste can take place; this is particularly so if artificial liners are to be emplaced (see paragraph 4.42). Such works may involve grading the base or sides of the site, the formation of embankments and the like, and require large tonnages of overburden or similar material to be moved. Material may

also have to be placed in stock-piles for later use at the site. All such operations will require the use of earth moving machinery. The number and type of machines needed will of course depend on the nature of the earthworks to be carried out.

4.39 Various features of landfill design may require substantial earthworks. The cell method of operation (see paragraph 5.16), which is being increasingly adopted at modern landfill sites, requires the construction of cell walls. This method of operation has the advantage that infiltration of rainfall is reduced and leachate generation is thus more readily controlled. The size of cells required will be influenced by the rate of waste deposition. It has been found that a 30 m wide cell can handle an input of some 1000 tonnes/day of household waste. It should be noted that when clay bunds are formed to a height greater than 4 m, specialist advice should be obtained on their design and construction.

4.40 At some sites earthworks may be needed to construct earth banks around part or all of the site perimeter to screen the landfill operations from the public. Trees or shrubs may then be planted on the banks to enhance the screening effect. Major earthworks may also be involved in the provision or improvement of access roads as well as the construction of haul roads within the site itself.

Numerous minor earthworks may be required. For example, the construction of drainage ditches and sumps in unfilled areas of the site for the separate collection of landfill leachate and uncontaminated rainwater.

4.41 A special form of site engineering is required at some coastal sites where deposition of waste is used as a means of reclaiming land. Bunds or sea-walls, which may need to be a considerable size, are built to appropriate designs, into the sea and the enclosed water pumped out before waste disposal begins. Smaller intermediate bunds may be used to sub-divide the enclosed area during land-filling operations.

Lining landfill sites

Introduction

4.42 The principal aim of lining (and capping) a site is to contain the leachate, thus preventing pollution of surrounding land and waters (see also paragraph 4.59). Lining may also assist in leachate control by reducing groundwater infiltration into the landfill; particular applications include landfilling below the level of the water table, or the reclamation of low-lying ground in coastal areas where inflow of groundwater occurs. Lined landfill sites can usually accept a wider range of wastes than would otherwise be possible. Artificial liners are constructed of materials which are to all intents and purposes impermeable. Natural lining materials, such as heavy clay soils, exhibit low permeability (for example 10^{-8}cm/sec). As well as impeding the flow of liquid, natural liners may also attenuate the leachate.

Site preparation for liners

4.43 Where the use of a liner is envisaged, the suitability of a site for lining will have been evaluated in the site investigation stage and if found to be suitable, plans for its

lining will have formed part of the working plan for the landfill operation. Table 4.43 lists some of the factors that should be taken into account when assessing the suitability of a potential landfill site for lining.

4.44 Liners should not be installed until the site has been properly prepared. The engineering of landfill sites for lining is a technically complex measure which should be entrusted only to competent operatives. Some of the work that may be necessary is described below.

Earthworks

4.45 The area to be lined should be free of objects likely to cause physical damage to the liner, such as vegetation including stumps and roots, hard objects, sharp rocks and the like. The area to be lined should be allowed to dry sufficiently to ensure that the surface can support men and equipment during installation of the liner. Particular care should be taken to deal with the consequences of rainfall during the operation. Any accumulation of water should be removed from the site without delay and installation allowed to proceed only when ground conditions are suitable.

4.46 Where a smooth base to the site cannot be obtained or where synthetic lining materials are being used, a blinding layer of suitable fine grained material should be laid to support the liner. A layer of similar material should also be laid above the liner to protect it from subsequent mechanical and environmental damage (see paragraph 4.64). Where synthetic liners are used, an 'anchor trench' around the periphery of the site to which the liner can be secured is usually required.

Under drainage

4.47 Whenever possible, sites in which a high groundwater table exists should be rejected as being unsuitable for lining. However, it is recognised that, occasionally, such sites may have to be lined, in which case an adequate

Table 4.43 Factors in the assessment of the suitability of a site for lining

1.	Grading of sides:	Side slopes of less than 1:3 are required to accommodate the plant and equipment used in lining; it must be possible to grade the sides of sites to meet this requirement.
2.	Hardness of base:	A base of hard rock will usually preclude the use of soil sealants such as bentonite which require rotovation of the site prior to installation.
3.	Stability of bedrock:	Lining materials are unable to cope with sudden or differential settlement.
4.	Underlying impermeable materials.	At some sites it may be possible to excavate the overburden to reach naturally imperme-able materials.
5.	Groundwater inflow:	The emergence of groundwater beneath either a natural or synthetic liner will weaken it.
6.	Base below water table:	In such cases the base will have to be raised with inert materials before the liner can be installed. Temporary lowering of groundwater by pumping is not recommended.
7.	Stability of subgrade:	If bund and cell walls cannot be engineered without fear of slippage there will be problems when laying the liner on slopes.
8.	Treatment or disposal of leachate:	Consider availability of facilities to deal with leachate in excess of site capacity.

underdrain system will be required. It should also be recognised that in the event of liner failure, any underdrain will provide a conduit along which leachate will readily migrate probably resulting in significant water pollution. If failure of a liner would, regardless of other precautions, result in unacceptable deterioration of water quality, such a site should be used only for inert wastes.

Provision for leachate management

4.48 At small landfill sites leachate collection is facilitated by grading the base of the site to a low point or sump from where it can be pumped away. Using permeable material as a protective cover for the liner allows the flow of leachate to the collection sump. As an alternative many lined sites have used French drains immediately above a liner but these are susceptible to blockage.

4.49 At large sites grading to a low point or using French drains is usually impractical. Spike drains have, however, been successfully used. These are trenches about 1 m x 1 m, excavated prior to lining which are then also lined along with the rest of the site. After lining, large perforated drainage pipes are installed in the trenches and backfilled with suitable coarse material. The layout is designed to provide a slight fall to a collection sump from where leachate can be pumped away for subsequent treatment and disposal. Leachate treatment and disposal options are discussed in paragraph 3.44 et seq.

Provision for monitoring

4.50 Lining landfill sites is primarily aimed at safeguarding groundwater quality. A continued groundwater monitoring programme, to confirm the integrity of the lining, is therefore essential. Such a monitoring programme should begin before installation of the liner to provide baseline data for comparison with later results. At an early stage of site preparation therefore, a number of monitoring boreholes need to be provided around the site. The location, design and number of boreholes should be agreed with the appropriate authorities and will depend on the size of the landfill, proximity to an aquifer, geology of the site and types of waste deposited. Borehole construction is discussed in paragraphs 3.113–3.115.

4.51 Leachate levels within a landfill should be monitored regularly to check that leachate is not building up to a point where water quality, site operations and restoration are put at risk. Various systems to monitor the level of leachate in landfills have been used, mostly based on pipes installed prior to landfilling. Small bore perforated plastic pipe is cheap and easy to install, but has the disadvantage that the pipes are easily damaged during infilling. Placing pipes within a column of tyres offers some protection. The use of stacked perforated concrete rings as infilling proceeds has been widely used and has the advantage that it can accommodate a suitably pro-

tected pump if leachate has to be removed. There may, however, be problems due, for example, to instability or leachate attack.

Liners constructed of naturally occurring materials

4.52 For a natural material to be suitable as a liner it should ideally have a permeability of less than 1×10^{-7} cm/sec. The most commonly used low permeability material is clay, either native to the site or imported. Alternatives include bentonite, colliery shale and Pulverised Fuel Ash (PFA). The use of these materials, including any disadvantages is discussed in Appendix 4a. The average permeability of any natural material proposed for use as a liner needs to be carefully evaluated by a competent person to determine its suitability. For example, the permeability of clays can be increased significantly by the presence of sand and gravel lenses.

4.53 Where natural materials are being reworked to a form suitable for use as an impermeable liner, the engineering properties of the material should also be carefully assessed particularly with respect to moisture content and ease of handling to provide good compaction. Also, since the liner will be exposed to leachate, the behaviour of the material in the presence of a range of organic and inorganic wastes should be determined prior to installation. Once emplaced, the permeability of a liner should be checked to ensure that it meets the design criteria, particularly in respect of the relationship between thickness of liner, permeability and hydraulic head of leachate expected.

Synthetic liners

4.54 Most commonly used synthetic liners are thermoplastics formed from polyethylene, polypropylene and polyvinyl chloride (PVC) either used individually or combined as copolymers. Additives may be incorporated to improve the properties of the material. Although many different copolymers and additives are available, in practice there are only a limited number suitable for use as lining materials. These are listed in Table 4.54 and discussed in Appendix 4b. Such materials come in various thicknesses and widths.

4.55 The service life of a synthetic liner is expected to be up to 30 years. However, since the use of liners is

Table 4.54 Synthetic lining materials

butyl rubber
chlorinated polyethylene (CPE)
chlorosulphonated polyethylene (CPSE or Hypalon)
ethylene copolymer bitumen (ECB)
ethylene propylene rubber (EPDM)
neoprene (chloroprene or polychloroprene)
polyethylene — low density (LDPE)
polyethylene — high density (HDPE)
polypropylene
polyvinyl chloride (PVC)

relatively new insufficient time has elapsed for data on their service life to be obtained. A summary of the results of laboratory testing is given in Table 4.55.

Table 4.55 Summary of findings from laboratory testing of synthetic liners

1. Liner material swelling and loss of strength was accelerated at high temperatures.

2. Not all liners within the same category had identical composition or physical properties as formulation can vary.

3. Liner compatibility with hazardous wastes is highly waste specific. Oily wastes had the greatest effect on liner properties.

4. Leachate caused liners to swell and lose strength.

5. Chlorinated polyethylene, chlorosulphonated polyethylene and neoprene showed the greatest changes in physical properties. Low and high density polyethylene performed well in all tests.

6. Seams were a weak point and several types of liner showed a loss in seam strength, particularly if adhesives were used. Heat sealed seams retained their properties well.

4.56 A reinforced bitumen liner has been available for some time. It can be tailored for a specific function by incorporating different types of reinforcing material in the membrane. To date, however, it has mainly been used as a liner for reservoirs where it has proved to be both durable and resistant to weathering. Its use for lining landfill sites is likely to be limited since bitumen is readily attacked by organic solvents and oils.

Choice of liner

4.57 The choice of a suitable liner material is a difficult decision since, not only are a large range of materials available, but their long-term behaviour is uncertain. No definitive recommendations can be given since the decision will inevitably depend on circumstances unique to the site. It is clearly important that the liner material selected must be compatible with the types of wastes to be deposited. It should be noted that several materials, notably butyl rubber, chlorinated polyethylene and chlorosulphonated polyethylene are weakened by contact with high concentrations of organic chemicals. Both low and high-density polyethylenes are compatible with the wide range of materials found at household refuse landfills. However, there is no definitive information available on their behaviour at landfills taking difficult wastes. Similarly, with the possible exception of some clays there is very little information on the behaviour of natural lining materials in contact with difficult wastes (see also paragraph 7.47).

4.58 The mechanical strength of a synthetic liner is also important. Damage to the liner must be guarded against at all times, by using as thick a material as practicable, taking into account the difficulties in handling synthetic liner

materials. It should be borne in mind that regardless of the protection afforded to liners their integrity cannot be guaranteed at sites where significant settlement of the subgrade could occur.

4.59 A synthetic liner material should also be resistant to weathering. During its placement some of the material inevitably will be left uncovered, or only partially covered; it should therefore be able to withstand exposure for at least six months without significant deterioration taking place. The material should also be resistant to attack by small mammals, particularly rodents. Similarly, it should be resistant to microbial attack and damage by plant roots. In this respect, natural materials may offer advantages over synthetic liners.

4.60 Finally the required service life of a liner should be considered since most manufacturers will guarantee properly installed synthetic liners and soil sealants for only up to 25 years.

Installing synthetic liners

4.61 Techniques for installing synthetic liners are well developed but remain highly specialised; improper installation remains the primary reason for failure of a synthetic liner. For this reason the experience and service offered by the supplier/installer should be carefully considered. Problems which may be encountered include:

(a) installation of damaged liner material,

(b) inadequate preparation of site,

(c) inadequate preparation of sheets for seaming,

(d) inappropriate sealing method used, and

(e) perforation of the liner by plant or machinery during installation.

Covering material for a synthetic liner

4.62 Cover is required to protect a synthetic liner from the effects of weather, from mechanical damage by birds, animals and vandals, and from damage during subsequent placement of refuse. Household refuse usually contains cans and glass and these as well as large items such as old refrigerators and furniture, can easily puncture synthetic liners. It is recommended therefore that a minimum thickness of 250-300 mm of cover material is used to safeguard against puncturing. On steep sided slopes it may be necessary to increase the thickness to achieve good compaction. While the material used for cover will usually depend on its availability, it should ideally be highly permeable to encourage drainage of surface water. Examples of waste materials which may be considered as suitable for cover material and qualifications to their use are given in Table 4.62.

Table 4.62 Examples of wastes which may be suitable as cover for synthetic lining material

1.	Colliery spoil	High concentrations of pyrites may result in acidic drainage which adversely reacts with the liner. High sulphide or sulphate may prevent direct discharge to sewer.
2.	Limestone/chalk	Calcareous materials should not be used on sites where bentonite has also been used as this will result in an increase in permeability.
3.	Coarse gravel	Large sharp stones may penetrate the liner.
4.	River silt	On slopes fine material may be eroded. Lower permeability than most other cover materials.
5.	Quarry fines	Difficult to handle when wet.
6.	Ground dolomite	Does not drain freely.

4.63 Care must be taken to ensure that material laid above or below a synthetic liner is well compacted. It is also necessary to ensure that the cover material has not been eroded before the first layer of refuse is placed. Steep slopes should therefore be avoided. It is advisable to prepare and line only small sections of a site at any one time. Planning a lining operation should be undertaken with extreme care.

Permeable liners

4.64 The use of permeable liners to provide a zone of attenuation beneath a landfill is a new concept which is currently being explored. The aim is to enhance the attenuation capabilities of materials either occurring at the site or readily available nearby so that leachate movement can be delayed and attenuation mechanisms given a longer time to act before the leachate migrates from the site. Preliminary results are encouraging, but it will take some time to establish whether this technique is acceptable for general use.

Operational considerations for artificially lined sites

4.65 The operation of landfills is considered generally in Chapter 5. However, some special considerations that apply to the operation of lined landfill sites are highlighted below.

4.66 During the early phase of operation, particular care must be taken to ensure that traffic does not damage the liner. Traffic movements over cover material should be minimised and hardcore roads underlain by heavy duty fabric should be provided to permit vehicles to unload and turn. Particular care should be taken in placing the first lift of refuse, and build up of water and leachate should also be controlled. Where a drainage system has been installed above the liner, pipes should be cleaned out after the first lift of refuse. Subsequently cleaning annually is recommended.

4.67 One problem unique to landfills lined with polymeric materials is that should leachate have penetrated the liner it may then undergo further degradation and produce landfill gas. Gas build up has been known to result in billowing up of the liner. Some provision for gas venting may therefore be desirable by using a permeable substrate on which the liner is placed.

4.68 The performance of a liner should be closely followed by monitoring the quality of groundwater close to the site. However, it must always be borne in mind that in the event of a leak occurring, pinpointing its origin and repairing the damage is almost impossible. One option to minimise further groundwater pollution is to ensure that the level of leachate within the site is kept to a minimum and to cap the site with an impermeable material as soon as possible. Even then it may be necessary to continue to pump out leachate at regular intervals.

4.69 A more detailed discussion on various synthetic liner materials is given in Appendix 4b.

Preparation for leachate and landfill gas management

Leachate

4.70 Leachate management will have been considered as part of site assessment and suitable measures will have been incorporated in the working plan. Preparation of the site may be required at an early stage of development to divert surface or groundwater inputs and so to minimise leachate production.

Surface water

4.71 Preparatory work to control surface waters may include the diversion of existing watercourses, provisions to collect and disperse surface water run-off from adjacent land and facilities to deal with surface water arising on the site. Water from streams or ditches adjacent to the site may be diverted by constructing new channels, ditches or culverts, or by sealing barriers. A peripheral ditch system around the whole or parts of the site may provide an inexpensive solution. For shallow gradients wider channels or ditches will be required to accommodate the expected maximum flow; it is reported that gradients of less than 1 in 400 are normally unsatisfactory. Where a ditch is not likely to act also as a soakaway, additional soakaways may be required, or, where one exists, the system may need to be connected to a surface water drainage system. Weirs, bunds, small dams or penstocks may also be necessary, particularly in areas of high water table or where the land is prone to flooding.

Groundwater

4.72 Groundwater ingress will have been assessed and

account taken of its likely effects on leachate management as part of the working plan. The effect of hydrostatic pressure beneath a lined site also will have been considered and the weight of cover required to counteract any lifting effect, until the waste deposits provide sufficient weight, will have been calculated. A peripheral excavation down to impervious strata around the site, backfilled with a sealing material such as clay or bentonite, may provide a successful, albeit relatively expensive, method of restricting groundwater ingress.

4.73 Lowering the water table by pumping from strategically placed boreholes can be effective during the early stages of landfilling wastes. The consequences of ceasing to pump groundwater and the restoration of natural groundwater flows, particularly in relation to the types and quantities of waste deposited and their pollution potential, will have been considered as part of the Working Plan. The use of this procedure is likely to be worthwhile only to protect a landfill liner against damage by hydrostatic pressure beneath the lining, and then only if the consequences of liner failure would present a serious pollution threat.

Leachate collection

4.74 At containment sites a sump or sumps will be required to collect leachate for subsequent treatment and discharge. Leachate collection sumps will require facilities to pump leachate from the sumps either to a treatment facility or sewer. In all cases the sump should be provided on secure foundations and the structure designed such that its depth can be increased as waste deposits rise around it and withstand lateral forces which will inevitably act upon it.

4.75 Where waste deposits will eventually rise above the level of surrounding land an inner ditch system may be necessary to collect leachate moving laterally and to intercept surface water run off. Wherever possible arrangements for surface water collection should be kept separated from those for leachate. A system of weirs, penstocks, sumps and pumps may be necessary to collect leachate for transfer to a treatment facility or discharge to sewer.

4.76 Where leachate is to be treated before removal from the site, tanks or lagoons will be required for holding, treating and acting as predischarge buffer storage. These may simply be sumps at convenient low points in the site, or specifically designed tanks or lagoons with synthetic liners. A system of pipework and pumps therefore will be required to transfer leachate. Mobile pumps and semi-rigid hoses allow flexibility of operation.

4.77 The water authority will have been consulted and their agreement received before undertaking any activity affecting or likely to affect surface and underground waters and before any discharges of leachate take place.

Landfill gas

4.78 Where the nature, quantity and depth of the wastes to be deposited suggests that significant quantities of landfill gas will be produced at an early stage of the development, preparation will be required for its eventual handling and control. The principal design consideration will have been to prevent the uncontrolled escape of gas and thereby minimise fire, explosion or asphyxiation risks. Necessary measures may involve the provision of gas migration barriers and adequate ventilation and gas collection wells. The control and use of landfill gas is considered in detail in Annex 2.

NATURAL LINING MATERIALS (see paragraphs 4.52 and 4.53)

Clay

1. Clay when used as a lining material has properties that can minimise leachate migration, and also attenuate polluting species.

2. Its main disadvantage is that it is difficult to handle when too wet, it is difficult to spread properly and when too dry, becomes hard and cannot be compacted. Care must also be taken to ensure that the clay has not been weathered, since weathered clay compromises the integrity of a liner. Saturated rehandled clay has poor stability characteristics.

3. It is unlikely that leachate derived from household wastes will have any significant effect on the permeability of clay. However, interactions between clay and industrial wastes, particularly those containing certain organic materials, may result in increased permeability due to dehydration and hence shrinkage of the clay. Alkaline inorganic wastes can also react with clay and increase its permeability. Lime rich clays similarly will be rendered more permeable by acid attack. Where large quantities of acidic and/or organic wastes are to be deposited the properties of the clay used as a liner should be evaluated.

Bentonite liners

4. Commercially available bentonite clays are mainly sodium or calcium montmorillonite. In addition, preparations containing polymer additions to bentonite are available for uses where the fluid to be contained has a dissolved salt content greater than 1,000 mg/l. For most landfill operations bentonite with polymer addition will usually be required.

5. When mixed into soils and wetted, bentonite can swell up to 15 times its initial volume, thereby filling voids and creating an impermeable barrier. It has been shown that even when mixed with sandy soils bentonite can provide a barrier with a permeability of less than 1×10^{-7} cm/sec.

Colliery shale

6. Colliery shale has been used both as a material to protect liners and as a liner in its own right. Its suitability as an impermeable liner depends on its ability to be compacted. With good compaction a permeability as low as 1×10^{-7} cm/sec can be achieved. Its main disadvantage is that, particularly when left exposed after placement, it is degraded by weathering. Apart from increased permeability, weathering may initiate chemical reactions which produce acidic leachate high in sulphide or sulphate. To minimise such adverse affects the use of already weathered material from spoil heaps where leaching has already taken place should be used. Although colliery shale has been used at several household waste landfill sites, its long term effectiveness is unknown.

Pulverised Fuel Ash (PFA)

7. The suggested use of pulverised fuel ash as a lining material for landfill sites has been based on its permeability and pozzolanic properties. On lagooning, the permeability of PFA is approximately 10^{-3} cm/sec and following placement and compaction can reach 10^{-5} cm/sec. Its pozzolanic properties result from hydration reaction which cannot be controlled or predicted with any degree of certainty and in any case varies with different origins of the coal originally burnt.

8. Unless conditioned by lagooning, soluble boron compounds in PFA will appear in run-off water. In addition, dissolution of other elements including magnesium, sulphur, manganese, vanadium, and chromium by leachate from deposited waste, can occur. Attenuation of potentially polluting species present in waste is not predictable since these properties also vary with the origin of the coal originally burnt. At the present time the use of PFA as a lining material for landfill sites to minimise leachate migration cannot be recommended.

SYNTHETIC LINING MATERIALS (see paragraph 4.54)

Butyl rubber

1. Butyl rubber was first used to line reservoirs more than 25 years ago. It is a thermoset plastic and is available as either unsupported or fabric reinforced sheeting in thicknesses of 0.51 to 0.63 mm (20 to 25 mil). It has excellent resistance to swelling and permeation by water.

2. Its use as a landfill liner is limited, since, while not affected by mineral acids it is susceptible to attack by hydrocarbon solvents and oils, even when they are present in trace concentrations. Since both industrial and household wastes are likely to contain oils and similar materials, its use as a liner for landfill sites is not recommended.

Chlorinated polyethylene

3. Chlorinated polyethylene is highly resistant to deterioration in a landfill environment. It displays good resistance to microbial growth and remains unaffected when left exposed to the elements. However, it is not compatible with aromatic hydrocarbons and oils. Results from studies in the USA suggest that it may absorb leachate. Chlorinated polyethylene has not so far as is known been used as a landfill liner in the UK.

Chlorosulphonated polyethylene

4. Chlorosulphonated polyethylene commonly known by its trade name 'Hypalon', is available in various grades in both thermoplastic and vulcanised compositions. Sheeting is normally reinforced with a polyester or nylon scrim to improve tear strength and puncture resistance. It is easy to seam and displays good resistance to microbial growth. Unfortunately it has been demonstrated to offer poor resistance to weathering. Exposure for even a short time to sunlight can cause shrinkage and hardening. Chlorosulphonated polyethylene liners also have poor resistance to hydrocarbons. They have not so far been used to any extent in the UK, but have been used in several other European countries.

Ethylene copolymer bitumen

5. Ethylene copolymer bitumen is a thermoplastic producted in several grades including sheets designed specifically for landfill lining. The product has only recently been marketed in the UK, but it has been used extensively in Germany for the past 10 years. Published information on its field performance is limited but it is understood to be resistant to attack by leachate.

Ethylene-propylene rubber

6. Ethylene-propylene rubber liners have not been used widely. Laboratory testing has shown them to be tolerant to a wide range of environmental extremes as well as being resistant to dilute acids, alkalis, silicates and brines. However, contact with hydrocarbon solvents or oil causes weakening of the liner. No information on the field performance of this lining material is available.

Neoprene

7. Neoprene is similar to natural rubber in flexibility and strength, but has improved resistance to weathering and mechanical damage. Unlike other materials, neoprene is resistant to dilute solutions of hydrocarbons and oils. Its main disadvantage as a landfill liner is its high cost and the need to use adhesives for seaming.

Polyethylene

8. Two types of polyethylene liner are currently available:

   (a) low-density polyethylene (LDPE) and

   (b) high-density polyethylene (HDPE).

The major distinction between the two materials is that low-density sheet is available only with a limited thickness, while high-density polyethylene has recently been introduced with a thickness of up to 3.2 mm. The thinner low-density polyethylene sheets are more difficult to handle in the field, and offer poor resistance to puncture and mechanical damage. The greater thickness of the high-density polyethylene liner on the other hand offers increased strength and resistance to puncture and abrasion. It is more capable to withstand stresses caused by moderate settlement of refuse or subgrade. The disadvantage in using the increased thickness of high-density polyethylene is its limited flexibility. It is also liable to thinning (creep) when subject to stress over an extended period of time. Both low and high-density polyethylene can be seamed by any of the available methods. Both materials offer very good resistance to a wide range of industrial chemicals, including oils and solvents. Both materials perform well in laboratory and field tests. Low and high-density polyethylene are the two materials most widely used for landfill lining in the UK. Some liners, using these materials, have been in use for nearly ten years and no information has come to light to indicate that these materials do not function effectively.

Polypropylene

9. Polypropylene offers properties which make it suitable for use as a landfill liner under a wide range of conditions.

Polyvinyl chloride (PVC)

10. Plasticised polyvinyl chloride is widely used for lining waste treatment lagoons. It is available as unsupported sheeting up to 0.75 mm thick, having good tensile strength properties and resistance to mechanical damage. It is also resistant to many inorganic chemicals, but is weakened by hydrocarbons, solvents and oils. Its useful life is limited to about 20 years since the plasticiser that forms up to 40% of its composition is slowly leached out or volatilised. Loss of plasticiser causes the liner to become brittle and therefore susceptible to stress cracking. Polyvinyl chloride has not been used for landfill liner applications in the UK.

60

CHAPTER 5: THE OPERATION OF LANDFILL SITES

Introduction

5.1 All landfill operations require careful planning in advance of the first deposit of waste; indeed, the main features of the proposed operations will have been set out in the site operational plan which forms part of the working plan (see paragraph 3.120). How a landfill is operated determines to a large extent the environmental effects, and hence the public acceptability, of the operation. Good landfill is also a major factor in relation to the economics of the operation. For example, there is every incentive to extend the operating life of a site for as long as possible, the more so since landfill sites generally are not easy to acquire. This can be accomplished by ensuring that the incoming waste is deposited so that void space is used in an efficient manner consistent with other operational requirements of the site. Such a requirement has a vital influence on the mode of site operation (see paragraph 5.28).

5.2 One basic factor influencing the planning of site operations is the nature and quantity of incoming waste. For example, the type of filling operations to be undertaken will be influenced by whether the waste is crude or pulverised or baled, household or commercial or industrial, alone or mixed. This in turn will affect the type of machinery required, and also the use of cover material and litter control. The type of delivery vehicle also needs to be considered as this may influence site road requirements and unloading areas. The anticipated rate of waste deposition should have been determined prior to the start of landfilling. Consideration should have been given to not only the rates currently anticipated but also to predicted rates over the expected life of the site, taking into account, for example, any increase resulting from a redirection of waste from other sites nearing completion. These data are important in terms of planning site restoration, controlling leachate generation, staffing levels and selecting machinery. It is important that the plant on site is capable of undertaking the required tasks and is efficiently utilised.

5.3 The application of cover material during or at the end of the working day is an essential part of landfill practice. The type of cover material used and its availability are important aspects when considering how the site will be operated.

5.4 The various factors that are important to the efficient and responsible operation of a landfill are considered in this chapter. It should be noted that certain special considerations that apply to the operation of sites taking difficult wastes are dealt with in Chapter 7.

General operational factors

Pretreatment of household and commercial waste

5.5 Following its collection household waste will usually be taken directly to a landfill. However, in some cases pretreatment has been found to be advantageous. The most commonly used pre-treatment methods for wastes in the UK are baling, wet pulverising and dry pulverising.

5.6 Various types of baling machine are available which can compress domestic waste to densities ranging from 0.75 tonnes/m^3 to over 1 tonnes/m^3. The size and shape of the bale obtained depends on the design of the machine used; bales weighing at least 1 tonne are commonly produced and these are either cohesive or need to be wire bound. The baling of waste facilitates its transport to a landfill site where the bales can be mechanically handled and neatly stacked, thus giving the site a tidy appearance, especially since very little litter is produced (see also paragraph 5.26).

5.7 Wet pulverising of domestic waste has been used to give a more homogeneous material which, it is claimed by some, can be deposited in a landfill without the need for the use of daily cover material. Crude household waste is fed into an inclined rotating drum together with the appropriate quantity of water. A residence time of several hours is usually required after which time it is generally found that 50-80% of the input material is pulverised. It is normal practice to landfill any rejected material separately. However, it should be noted that wet pulverised refuse possesses a much lower water absorptive capacity than crude refuse. This fact should be taken into account when calculating the water balance (see paragraph 3.33 et seq).

5.8 In most dry pulverisers, waste is reduced in size by the use of rotating hammers, usually after the separation of ferrous materials by electromagnets. The process produces a dusty material which can be landfilled but is sometimes used as feedstock for various reclamation or fuel making processes.

Waste reception

5.9 The importance of maintaining a tidy site entrance is referred to in paragraph 4.4 and indeed such a requirement is likely to be a condition of the disposal licence. Landscaping, including the planting of trees and bushes may be a planning requirement. Where the site entrance includes a reception area for waste brought by the public special care must be taken that this area does not become an eyesore (see paragraph 4.29).

5.10 The types and quantity of all incoming waste should be recorded to provide data for a continuing assessment of waste inputs and with cover requirements as well as enabling predictions to be made on rate of filling, engineering requirements and future restoration. Depending on circumstances the rate of filling may be determined from the records, aided by the installation of a weighbridge (see paragraphs 4.13-4.17), or measured by regular surveys of the site void. It is important that frequent inspections are made by both the operator and the WDA to ensure that wastes deposited are only those permitted by the disposal licence.

5.11 Those delivering waste to the site should be made aware of the site regulations, for which at least a copy of the following documents should be displayed in a prominent location:

(a) the current waste disposal licence,

(b) the employers liability certificate,

(c) the terms and conditions of trading, and

(d) the site regulations including a traffic plan.

Other site notice board requirements are described in paragraphs 4.27 and 4.28.

5.12 The control of traffic is very important, particularly on a large site where the working face may be a long way from the site entrance. The route should be clearly sign-posted (see also paragraph 4.28). Where difficult wastes are accepted it is essential to ensure that loads are deposited under close supervision. Children should not be allowed to enter landfill sites and adult passengers should be required to remain in the cabs of vehicles.

Preparation and maintenance of site roads

5.13 The importance of good quality site roads, and considerations that should be borne in mind in designing the site road system, have been discussed in Chapter 4 (paragraphs 4.30-4.32). Good quality site roads are essential to ensure a swift turn-round of vehicles in all weathers. Good site roads may also remove the need for an emergency tipping area. Road maintenance should ideally be carried out during the summer months. Good planning and its implementation is essential to ensure that site roads are located where they are required.

5.14 Temporary roads across deposited waste should be constructed of materials which will not cause damage to vehicles but will nevertheless afford effective traction across the site. For example, wastes such as builders rubble, macadam wastes from road repairs and the like should be stockpiled for use when required. The road surface should be raised at least 0.3 m above the surrounding deposited waste to allow water to be shed rapidly. The hardcore may be contained by putting down a fabric

membrane the underside of which is protected by sand. This can then be surcharged with a further layer of stone. In this way a reasonable road can be prepared which may be topped with macadam if necessary. Another system which has been used successfully involves the use of interlocking concrete slabs. The slabs are usually provided with lifting hooks so that they may be moved and added to as the landfill face advances.

5.15 Steel wheeled compactors must not be run over site roads; separate roadways for compactors should be provided.

Method of filling

5.16 Three variations in landfilling techniques can be distinguished although in practice, depending on details of site operations and conditions, the distinctions between them may become blurred. The three main techniques are as follows:

(a) *Trench method* This involves the excavation of a trench (which may be very large) into which waste is deposited. The excavated material is then used as cover. This technique is a variation of the cell method described below. It should be distinguished from the construction of trenches into already deposited solid waste for the disposal of liquid wastes. The trench method has found very limited application in the UK.

(b) *Area method* Waste may be deposited in layers and to form terraces over the available area. However, with this type of operation, excessive leachate generation may occur unless high waste inputs are maintained thereby providing adequate absorptive capacity to account for rainfall. This method has been used widely in the UK but is no longer favoured since operational control may be difficult.

(c) *Cell method* This method involves the deposition of waste within pre-constructed bunded areas (Figures 5.16A and B). It is now the preferred method since it encourages the concept of progressive filling and restoration. It is a method which is beginning to have widespread application and is accordingly described in detail below.

The cell method

5.17 Daily cells are often constructed within a larger cell. The sequence of landfilling will depend on many factors such as topography, rainfall, traffic flow, and method of deposition. At a shallow site it is considered preferable to place one daily cell on top of another so that the larger cell is brought up to final level before moving onto the next larger cell thereby assuring progressive restoration. However, in a deep site having a relatively small area such procedures may not be practicable and it may therefore be necessary to operate a cellular system of landfilling moving progressively across the site before moving onto

Figure 5.16(a)   METHOD OF LANDFILL OPERATION

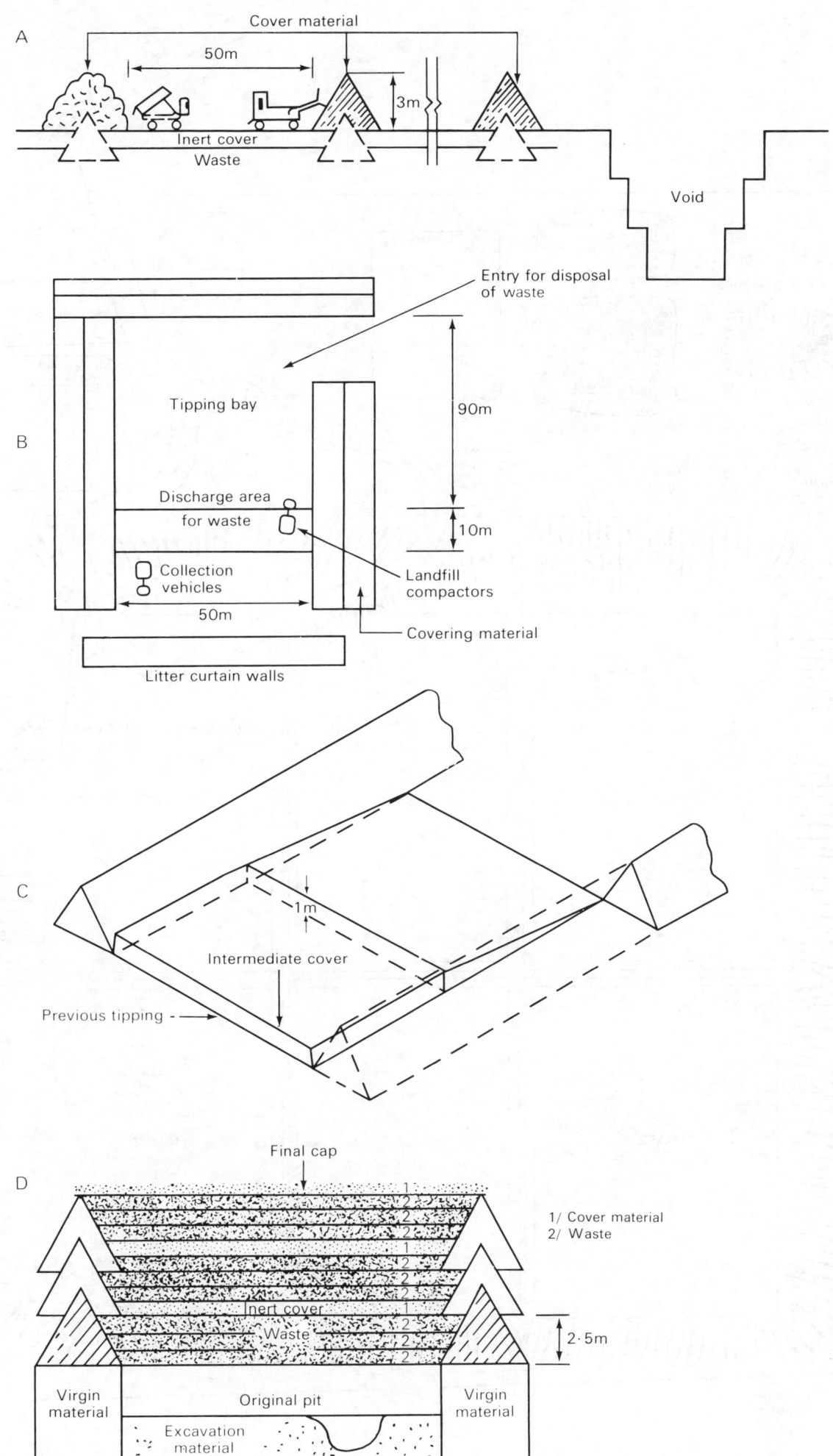

Figure 5.16(b) TYPICAL OPERATIONAL PLAN FOR LANDFILL SITE

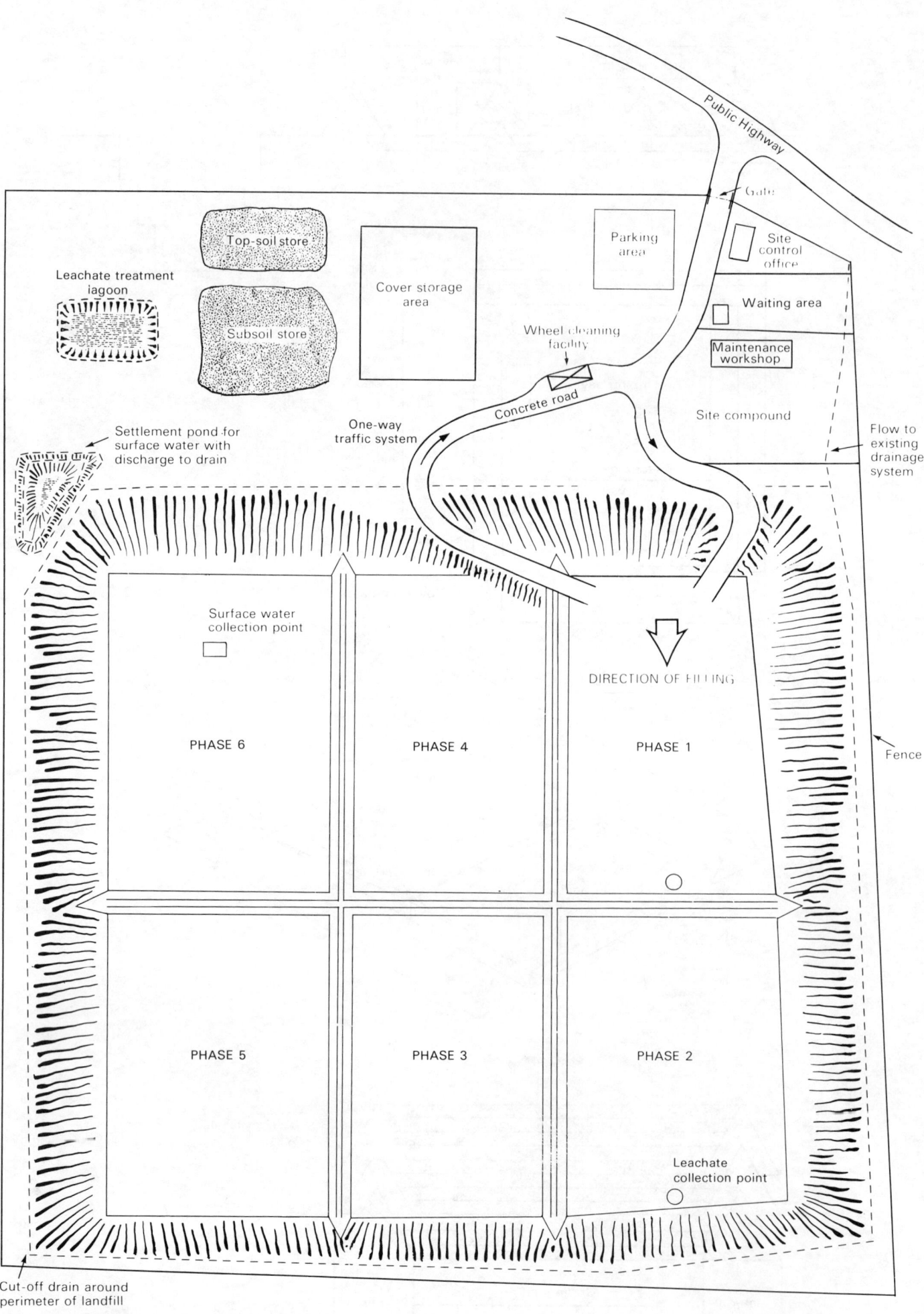

the next lift. However, the resulting large open area of waste will permit significant rainfall infiltration unless large quantities of cover material are used. High rates of waste input therefore offer advantages for deep sites. Operating a cellular method of filling enables waste to be deposited in a tidy manner since the bunds serve to both conceal the tipping operation and at the same time trap much of the litter which may be generated. When a cell is brought up to its final level, it can be capped thereby minimising infiltration of rainfall and consequent problems of leachate control. A disadvantage of the cell method, particularly where space is limited, is the amount of void space which may be lost due to building the cell walls. This drawback may be overcome in two ways: by utilising suitable incoming wastes for wall construction or by removing the wall at the end of the day and using the material as cover.

Cell construction

5.18   The size of cells should be influenced by:

   (a)   rainfall,

   (b)   absorptive capacity of the solid waste,

   (c)   the rate of input,

   (d)   number of incoming vehicles,

   (e)   sufficient space for safe turn-round of vehicles.

By careful design and operational management it is possible to ensure that the liquid absorptive capacity of deposited waste is not exceeded and thus minimise the generation of leachate. This is very important since particularly at containment sites excessive leachate generation may lead to treatment problems. Also it has the advantage that excess absorptive capacity can be retained within the site which may be utilised in the event of leachate build up caused by water ingress, for example, following unusually heavy or prolonged rainfall, or later, to accept liquid wastes as a means of their disposal.

5.19   Cell walls should be at least 2–3 m higher than the height of the daily lift. One end of the cell should be close to a secondary access road. Initially cell walls may be formed by pushing up material from the base of the site taking care not significantly to reduce the depth of any impermeable layer protecting water resources beneath the site. Subsequently it will be necessary to raise the walls as filling proceeds. This can be achieved by using suitable waste materials. It is important to ensure the structural stability of all walls (see paragraph 4.39). Ideally, low permeability material should be used in order to prevent leachate seepage through the walls which could contaminate clean surface water. Where there is a risk of groundwater ingress, low permeability cell walls also reduce hydraulic continuity across the site. A disadvantage is that cells constructed of low permeability material may

encourage the build-up of leachate at different heights within the site. Where possible, a cell (or cells) should be taken up to final level by continuous filling but usually the use of intermediate cover will be necessary. The use of a low permeability intermediate cover material, such as clay, helps to reduce infiltration; however, any temporary cap using such material should be scraped off or penetrated before subsequent layers of fill are placed to avoid the build-up of perched water. This is particularly important where waste disposal above-ground level is planned, since perched water could then result in surface discharges. Infiltration can also be minimised by capping and sloping the surface of a completed cell and at the same time channelling or pumping run-off away from the operational area.

5.20   The working face should be sufficiently extensive to permit vehicles to manoeuvre and unload quickly and safely without impeding refuse spreading and compaction and allow site equipment to be operated easily. A balance must therefore be achieved in relation to the number of incoming vehicles and the need to provide cells sized to minimise infiltration, cover requirements and litter. The size of a working face will require regular review as operations proceed.

Refuse placement

Deposition of crude refuse

5.21   Where the base of the site cannot support the weight of vehicles, a preformed base will be required. Pockets of silt or sludge can present special operational problems. In many instances it is preferable to use equipment that has a low ground pressure, such as a wide tracked dozer, for the first lift. The disposal of waste, other than inert materials, into standing water should not be permitted.

5.22   Pushing waste over a vertical face is not acceptable. It should be deposited at the top or base of a shallow sloping working face. Machines may then be used to push the waste up or down the slope. The method predominant in the UK is to deposit waste from the top of a working face. A high degree of compaction can then be achieved, provided that the operator spreads the waste thinly. By using a shallow working face any risk that vehicles depositing waste may reverse over the edge is eliminated. Conversely waste may be deposited at the base of a working face and be pushed up the slope in thin layers. This has the advantage that since only a modest quantity of waste can be handled in one push, there is less tendency to make layers too thick. A further advantage is that because the waste is being discharged from vehicles at a lower level than the rest of the site, windblown litter is less likely (see paragraphs 5.56-5.60).

5.23   The angle of a working face should be shallow.

Operating with a face slope greater than 1 in 3 causes considerable problems for delivery vehicles and compaction plant. A suggested optimum slope is about 1 in 12. Other advantages of using a gentle slope are reductions in accidents, improved fuel consumption and reduced wear and tear on the machine, while at the same time the waste density is increased.

5.24 Depositing waste in thin layers and using a compactor enables a high waste density to be achieved. Each progressive layer should not be more than 0.3 m thick. The number of passes by a machine over the waste to achieve optimum compaction will depend on a number of factors including the type of machine, its ground pressure, type of waste and slope. Obviously, the more passes made over a waste, the better its compaction, but operational considerations generally limit the number to between two and six. It should be noted that some of the most awkward wastes to handle are dry cardboard, paper and plastics which can be readily windblown. To compact such materials more passes may be required.

Deposition of pre-treated wastes

5.25 Pulverised wastes can be landfilled in much the same manner as crude refuse. Rejects, which may comprise up to 50% of the incoming waste, can be deposited first and then covered with the fine material. Daily cover may still be required.

5.26 Baled waste can be transported to the working face on a trailer and should be deposited with the aid of a fork lift truck specifically designed for use over rough terrain. Conventional forklift trucks can be operated only where there is a firm, level and preferably concreted base. Careful handling of the bales is essential as experience shows that up to 30% may break open if handled more than once. In practice a stack height of three bales per lift is recommended; to lift less than this is considered uneconomic, whereas with four or more bales per lift it is difficult to position them correctly. In addition unloading time is increased and there is a safety problem when stacking higher than a fork lift truck. To provide stability to the lift, the bales should be laid in a similar fashion to bonding brickwork rather than directly in line on top of each other. This also has the advantage that in the event of fire there is less opportunity for it to spread along the gaps between the bales. Indeed, at sites taking baled wastes it is advisable that discrete cells are constructed and then infilled with the bales. The size of these cells should be restricted to minimise leachate generation and fire risks. Close stacking of bales is highly desirable for similar reasons.

Compaction

5.27 Compaction is essential for maintaining a well-run and visually acceptable site. It offers the following advantages:

(a) by increasing waste density, the life of the site is extended;

(b) a well and uniformly compacted layer of waste reduces the volume of daily cover required;

(c) a well compacted site is visually more acceptable and there is less risk of litter blowing across the site;

(d) compaction reduces colonization by vermin and fly infestations;

(e) by eliminating voids, underground fires are largely prevented and surface fires are made much easier to control;

(f) well compacted waste provides a more stable base for delivery vehicles during discharge of loads. This reduces vehicle wear and tear and the risk of machinery becoming bogged down during wet weather; and

(g) a high degree of compaction reduces settlement while encouraging more even settlement.

5.28 A consequence of compacting waste to a high density is that biodegradation may be slowed down due to decreased infiltration of water. Consequently onset of production and the period of landfill gas and leachate generation may be extended.

Cover materials

Covering of waste

5.29 At most sites at the end of the working day, all exposed surfaces, including the flanks and working face, should be covered with a suitable inert material to a depth of not less than 0.15 m. This 'primary' or 'daily cover' is considered essential as:

(a) it improves the appearance of the site and minimises windblown litter;

(b) depending on the type of material used, movement of vehicles over the waste can be facilitated;

(c) it will help to reduce landfill odours;

(d) it will inhibit colonisation of the site by rodents and flies;

(e) it may help to control infiltration of rainwater into the waste; and

(f) it will minimise the risk of fire.

5.30 Daily cover fulfills only transient function and

should not be confused therefore with intermediate or final cover. Intermediate cover refers to material used for the end of a phase of landfilling whilst final cover refers to the material used for covering of the final lift of deposited waste.

5.31 There may be circumstances where restricted use of daily cover may be acceptable, such as at sites where windblown litter is not a problem. However, at most sites daily cover is essential to the proper running of the site.

Sources of cover

5.32 Cover may be obtained by:

(a) good site husbandry and efficient stockpiling of inert waste materials coming onto the site,

(b) on site excavations where geological conditions permit,

(c) importing inert industrial wastes such as colliery spoil, pulverised fuel ash, and in some cases incinerator residues, and dredged silt or sand.

5.33 In addition the following alternatives may be considered on their merits although they are not widely used:

(a) artificial foam covering, although it is not suitable for windy or wet conditions and special safety precautions may be required,

(b) fines from pulveriser plants, but it has the disadvantage that it can contain a very high percentage of the fly larvae present in domestic waste and may also be odourous and thus increase odour emissions from the site,

(c) sewage sludge cake; this is not recommended as it may generate unpleasant odours and the sludge quickly reverts to a sticky mass when wet,

(d) cover may also be scraped from the previous layer and reused. However care has to be taken not to entrain previously deposited waste material.

Cover quantities and procedures

5.34 Whatever the cover material used, it will be necessary to estimate the quantity required each day. The theoretical amount required is in practice a minimum figure, as it is difficult to spread cover material to give an even layer. For example if fine grained material is used it will inevitably penetrate down into the waste rather than remain on the surface. As a result, the actual volume of cover needed to achieve the required standard could be as much as double the estimated quantity. It is important that the surface of the waste is rendered smooth and any large protruding objects are removed. Cover will still be required on sites receiving baled waste, although

the vertical faces may sometimes be left exposed.

5.35 Where a site relies on inert wastes for use as cover, it should be stockpiled at a point convenient to the working face. Cover should be applied throughout the working day. Spreading should be carried out progressively using an appropriate machine.

5.36 The choice of material for intermediate cover raises somewhat different considerations. The longer the time scale over which the cover is likely to need to be maintained, there will be greater emphasis on reducing infiltration, pointing to a need to use a low permeability material (see paragraph 5.18). However, the use of clay is not recommended for areas over which vehicles are likely to run, since it provides poor traction in wet weather. Pulverised fuel ash and mining dusts may present similar problems as well as giving rise to a dust nuisance in dry weather conditions.

Equipment, manning and safety

Equipment

5.37 The plant most commonly used on landfill sites includes steel wheeled compactors, tracked dozers and loaders, rubber tyred wheel loaders, scrapers and hydraulic excavators. However, auxillary equipment may also be required to carry out specific tasks such as spraying with water to reduce dust problems, the application of pesticides, roadsweeping and the like.

Steel wheel compactors

5.38 Steel wheel compactors have been developed for landfill operations during the last ten years and are now used extensively. Their cleated wheels are designed to break up and compact waste. At small sites receiving low volumes of waste, a compactor alone may be adequate to spread and compact the waste as well as handle and place cover material. However, a compactor is not designed to be a multipurpose machine and it is more usual to provide additional plant in the form of a tracked or wheeled bucket loader, particularly if cover has to be excavated or transported any distance. Properly trained operators, compacting waste in thin layers, enable compactors to be used in an efficient manner by offering the following advantages:

(a) they chop and compact waste;

(b) they give a high initial density;

(c) they produce an even and stable surface; and

(d) they 'pin down' waste thus minimising litter and make the site less attractive to birds and vermin.

5.39 Landfill compactors are available in a range of sizes from 100 to 300 Brake Horse Power (BHP) and operating weights between 15 and 30 tonnes. Apart from size, the more obvious differences between machines are the cleats on the wheels and the wheel configuration. The wheel configuration is relevant when determining the number of passes required to achieve the desired amount of compaction.

Tracked dozers and loaders

5.40 These are very versatile machines with sizes varying from 70-700 BHP although 100-200 BHP machines are normally used on landfill sites. Tracked machines can be fitted with a variety of blades or buckets. Correct selection of fittings is important and is determined by the duties required. For example, where large quantities of waste are handled a blade will be required whereas if cover material is to be won and spread then a bucket may be more suitable. The advantage of a tracked machine for landfill operations is that because of its low ground pressure, it can operate under most conditions particularly where the underlying ground is soft. It is also very stable and manoeuvrable when working over rough or sloping ground. One significant disadvantage is that it is not very effective at compacting waste due to its low ground pressure. However, by operating a sloping working face its effective ground pressure is increased. Track life can be adversely affected by wastes being handled.

5.41 A multipurpose bucket provides the ability to separate bulky items from incoming waste and place them where required. It also provides additional grading ability. A tracked dozer/loader is also capable of being used to maintain secondary roads, tow bowsers, extract trapped vehicles and a range of other site maintenance work.

Rubber tyred wheel loaders

5.42 These are versatile machines and therefore may be useful for small sites or as back-up machines on large sites. Tyred loaders have the advantage that they can travel on highways without risk of damaging the surface. Because of their greater speed they are more efficient than a tracked machine except in soft ground conditions. Punctures can be a serious problem but by providing tyre guards, using foam filled tyres and increasing tyre specifications the problem can be minimised.

Scrapers

5.43 These are used on landfill sites for excavating and moving cover materials in the course of site preparation and restoration. Essentially there are two types, tractor and box or motorised rubber tyred. Since these are single purpose machines they are normally found only on large sites (see also paragraph 6.45).

Hydraulic excavators

5.44 These are commonly used on landfill sites where their main function is excavation, mainly related to site preparation.

Auxiliary equipment

5.54 This includes tractors, dump trucks, slave vehicles (when waste is bought to the site in containers), fuel and water bowsers and also lighting sets, road brushes, pumps and pipework. Water bowsers are particularly useful during the summer for damping down dusts.

Equipment selection

5.46 No one machine is capable of performing all the tasks required to operate a landfill. When selecting plant, consideration should be given to a variety of factors including:

(a) *Site characteristics* Landfill sites vary and inevitably there will always be site specific factors which affect the type of equipment required for their operation. For example, particularly at containment sites pumps will be required to remove uncontaminated rainwater; separate pumping systems will be required to deal with leachate.

(b) *Site preparation* Where landfilling is being carried out in conjunction with minerals extraction, the operator may be filling one part of the site while preparing another. This may involve the excavation of material from the base either to form a cover or a liner, in which case either a scraper or hydraulic excavator will be required.

(c) *Quantity of waste* The number, type and size of machines required will clearly depend on the quantity of waste arriving at the site each day. The size of machinery selected should also be capable of handling any foreseeable increase in waste arisings.

(d) *Type of waste* The waste most easy to handle and emplace on a landfill site is baled waste using a rough terrain fork lift truck. Other wastes will need to be pushed, compacted and covered at the working area, accordingly at least a compactor or bladed machine will be needed.

(e) *Density of waste* In general, optimum use of void space requires wastes to be compacted to a high density. The highest density can be achieved most quickly using steel wheel compactors. Density values achieveable range from about 0.5 tonnes/m^3 using a tracked vehicle pushing waste from the top of a working face to at least 1 tonne/m^3 using a compactor working on thin layers of waste. The higher the density achieved the less is the risk of uneven settlement occurring.

(f) *Cover requirements* One factor influencing the selection of equipment is whether or not there is a need to win cover on site. For large operations where cover is to be won, a hydraulic excavator or drag line can be used in conjunction with dump trucks or a box scraper. When cover is carried only short distances, either a tracked or wheeled loader may be suitable. At sites where cover is delivered it may be possible to use the machine that is normally compacting waste.

(g) *Back-up requirements* The advantage of operating more than one machine is that if the machine used for spreading and compacting refuse is out of service, incoming waste can still be handled. Alternatively, contingency plans must be made either to direct waste to another site or to bring in a suitable machine. Some equipment manufacturers offer a repair or replace policy which is a valuable insurance against breakdowns. The reliability of equipment and availability of spares should always be considered when selecting equipment. It is most important not to undersize plant; apart from the problems of overworking that will then arise, operating standards will inevitably suffer.

(h) *Operator comfort and safety* It is desirable that the cabs of all machines used on landfill sites should be air conditioned and protected by appropriate dust filters. They should also be fitted with rear spotlights, rollover cab protection and audible reversing signals. It is also important that machine operators are properly trained in the safe operation of their machines and that adequate supervision is provided.

Inspection and maintenance of plant

5.47 Machinery breakdown can be costly not only in terms of repairs and possible replacement machine hire charges but also the problems that will be caused due to poorly compacted waste and absence of cover. It is therefore considered prudent to encourage daily and weekly inspection of machinery and practice preventative maintenance. Maintenance check-lists supplied by the machinery and plant manufacturer should be adhered to. Specific time should be allocated for maintenance, particularly for weekly inspections. Adequate facilities for maintenace must be provided in the form of garaging on site or be conveniently accessible nearby. Makeshift arrangements for major refits of heavy plant can lead to accidents.

Employment levels

5.48 Employees should be competent, well trained and adequately supervised; training should include site safety and first aid. Since a landfill site can pose dangers to both site operators and users, emergency plans should be laid down and tested from time to time. High visibility clothing should be worn at all times and all employees should be encouraged to be alert and safety conscious. Where visitors are allowed on site they should be closely supervised to ensure their safety.

Manning levels

5.49 The number of employees required to operate a landfill site effectively and safely will obviously depend on many factors. The principal considerations that need to be borne in mind are indicated below:

(a) Flexibility of both employees and plant is necessary to ensure that the number of operators and machines employed can be used effectively. A form of shift system may need to be operated to avoid down time during meal breaks. A back-up system will also be required to cover for sickness and holidays.

(b) A record of the quantity of incoming waste will be necessary.

(c) Where a large number of vehicles use the site, traffic control is important. A marshal controlling traffic flow and unloading may therefore be required. Two-way radio communication may be useful.

(d) If cover material is excavated on-site additional staff will probably be required.

(e) Litter control and frequent cleaning of fences and litter screens is essential. On a large or exposed site this may require additional man power.

(f) Duties such as road maintenance, environmental control and leachate management, will have to be accounted for in the staff budget.

(g) Where liquid wastes or other difficult industrial wastes are received, additional labour may be required to implement any special disposal procedures.

Safety

5.50 The management and workforce should be made fully aware of the existence of site safety regulations and the need to observe them at all times. Such regulations should also be brought to the attention of those using the site for the deposition of waste. Steep gradients and sharp curves on site access roads should be avoided. If this is not possible warning signs and crach barriers must be provided. Speed limits should be displayed and enforced by site operator.

5.51 Overhead power lines may cross the site. These should be either diverted or measures taken to ensure that the level of waste does not rise above a level agreed with the CEGB or local electricity board. At no time should vehicles or equipment be able to get within arcing distance of any power lines. The fact that tipper lorries and the tail gates of waste collection vehicles will be raised on site should be taken into account. The Health and Safety Executive has issued guidance on this matter.

5.52 The lack of stability, which is apparent with many large vehicles whilst discharging waste, needs to be taken into account. Where the surface of the landfill is unsuitable such vehicles should discharge their waste elsewhere. Totting by site operatives and delivery personnel should not be allowd (but see paragraph 5.57). It must be emphasised that the working area at a landfill site is dangerous since people and vehicles are manoeuvring in close proximity to each other.

5.53 Fire hazards exist at landfill sites, prominent 'no smoking' notices should therefore be displayed at appropriate locations on the site. Hard hats should be worn. Additional precautions may be necessary in quarries where there is a danger from falling rocks. Footwear having strengthened toe-caps and soles should be worn by all site operatives.

5.54 Trenches and lagoons used for liquid or sludge disposal should be fenced or be clearly marked with poles and bunting and each trench should be labelled to indicate the type of wastes being deposited (see paragraph 7.103 et seq). Measures should be taken to ensure that vehicles discharge only in the designated areas. Vehicles should not travel over unstable areas on a landfill surface.

Other considerations

Bulky wastes

5.55 At some landfill sites a large number of bulky items may require disposal. Where it is economically viable, metal items such as refrigerators and cookers may be removed and sold for scrap. All other wastes should be placed, where possible, at the base of the working face after crushing with for example, a loading bucket, thereby reducing void space. Drums containing liquids should not be landfilled (see paragraphs 7.118-7.125). Arisings of aerosol cans require due precautions to be taken against the possible risk of explosion; landfilling practice should reflect the hazards presented.

5.56 No bulky items, even after crushing, should be present in the first lift of refuse deposited in a site lined with a polymeric membrane due to the risk of damaging the liner (see paragraph 4.64). Similarly, bulky items should not be present in the final lift of waste in sites that are to be capped with a low permeability material since settlement of the refuse may result in large items piercing the cap. After-use of the land may also be adversely affected.

"Totting"

5.57 The separation, and removal for subsequent sale, of items such as scrap metal is known as 'totting'. Many site operators will not allow such a practice since they consider it to be both dangerous to personnel and to inhibit the efficient disposal of the waste. However, totting by permitted persons is allowed at some sites; it is claimed that the practice discourages unauthorised people from entering the site. If permitted, totters can perform a useful function in the removal of materials such as wire which could cause problems to site machinery. If totting is permitted then brightly coloured clothing should be worn by totters as a protective measure. Totting should not be permitted at or near to an active working face but should be carried out only in specially designated areas away from the working face. Loads which, on inspection, are found to contain material which can be salvaged can be diverted to the totting area where items suitable for recycling can be segregated.

Environmental control

Litter control

5.58 Litter control at a landfill is very important particularly at sites situated near to housing or recreation areas. While it may not be feasible to change the direction of a working face according to wind direction, Meteorological Office records of prevailing wind direction and strength should have been consulted when planning the filling sequence. One approach to minimise wind effects is to deposit waste at the base of a working face as in this way refuse is discharged from the vehicles at a lower level than the rest of the site. High bund walls will also help to trap litter. For them to be effective they should be built up as filling proceeds such that they are at least two metres higher than the previous lift. At exposed locations fencing along the top of bund walls may provide additional benefits. Operations using a minimum exposed working area also helps to control litter. Provision of permanent litter fencing around the perimeter of sites in exposed positions should have been considered as part of the operational plan. Mobile litter screens have been used with some degree of success but they must be located close to the working area. The screens must also be cleaned regularly.

5.59 Site operators should ensure as far as they are able, that all vehicles delivering waste to their site are in good repair and have adequately covered and secured loads. Correspondingly, drivers of empty vehicles leaving the site should ensure that all waste, especially lightweight plastic or paper, has been removed. It should be noted that there have been serious accidents resulting from the tail-gates of waste collection vehicles being inadvertently lowered on a crew member clearing out the residues left after discharging waste. The Health and Safety Executive has issued guidance on means of preventing such accidents.

5.60 Where waste contains a large proportion of paper and plastic, litter may be a particular problem but this can be minimised, where possible, by bagging such material at source. It is obviously necessary to lay and compact such waste as rapidly as possible. In windy conditions, it may also be necessary to add cover at the same time in order to 'knit' the waste. Likewise damping of some wastes is essential during dry weather.

5.61 Overhead netting of the filling area which may be specified in the disposal licence for bird control also limits the spread of litter. Both mobile and permanent net cages have been used but the latter restrict flexibility during the filling sequence. Mobile nets attached to a framework on rails have been used successfully at certain locations, but operational problems may be created by restricting working areas.

5.62 No matter which method of litter control is employed, its regular collection will always be necessary.

Bird control

5.63 Birds, in particular gulls and crows, can be attracted to landfill sites in large numbers, particularly where sites receive appreciable quantities of putrescible waste. This situation is undesirable since the birds may transfer pathogens to nearby reservoirs and crops as well as depositing excreta and food scraps over the neighbouring areas. Birds can also present a hazard if the landfill site is located near an airfield (see paragraphs 2.19, 2.21, 2.22).

5.64 Measures to control litter will minimise scavenging by birds. Additionally, other methods may be required which may include:

Birdscarers These are used on a number of sites but are only effective in preventing scavenging. Birds attracted to the site may still be a nuisance to aircraft.

Distress calls These discourage birds from visiting the site and are often used in conjunction with birdscarers and falcons. The calls must, however, be changed regularly as birds quickly learn to ignore them.

Falcons A number of operators now employ people to patrol the site with birds of prey. These are very effective.

Nets A high cost option is to hang a net above and around the filling area, such that birds cannot gain access to the refuse. This has been successfully demonstrated at a few sites. Permanent nets could be practical at a very deep site where the nets could be maintained

in position for a long period of time. Alternatively mobile nets might be used.

Pest control

5.65 As a result of improvements in landfill management, infestations of flies and rats are now less common. However, flies may still be a problem during the summer months. The problem is often more acute where waste is directed to the site via transfer stations, as in some cases several days may elapse between collection and deposition. In these circumstances eggs laid in putrescible waste may hatch en route. As the life cycle of the house fly can be completed in ten days in hot weather, there is obviously a serious risk of fly infestations. Animal carcasses and food wastes may also provide a breeding ground.

5.66 Good compaction and application of cover are essential to the control of flies and rats. In circumstances where a minimal amount of cover is used, fly control can be achieved by ensuring that the surface area of waste is covered by the next lift of waste within six days. Fly maggots may, however, penetrate the cover material and in such situations the application of insecticides may be necessary. One approach adopted on a number of sites is to combine a long term insecticide with one capable of offering a quick knockdown effect. Special care must be taken by the operators when handling insecticides. They should be informed of the nature of the materials they are handling, instructed in its use and should be provided with the appropriate equipment and protective clothing. Great care must be taken to ensure that insecticides do not appear in leachate or pose airborne pollution problems (see MAFF Code of Good Agricultural Practice ref BL 5160. 1985).

5.67 The most satisfactory way to counter rat infestation is by effective site management. It is also desirable to arrange a system of regular visits and precautionary action by the local authority rodent control officer or a pest control contractor. Clearance of rats prior to commencing operations is recommended.

5.68 Under the provision of the Weeds Act 1959 the Ministry of Agriculture can require the occupier of any land to take the necessary action to prevent the spread of injurious weeds such as spear thistle, field thistle, docks and ragwort. (But see also Schedule 8, Wildlife and Countryside Act 1981.) A weed-free corridor around the site perimeter may be required.

Odours

5.69 Objectional odours from landfills can be significantly reduced by good site management. Recent research indicates that odour is likely to be a nuisance rather than a health hazard provided that there is sufficient dilution of the landfill gas to ensure that the concentration of methane does not build up to its lower explosive limit (5%

in air). The characteristic landfill odour is generated by the decomposition of the putrescible matter in household waste. Where waste has been directed to a site via a transfer station, it may give rise to a stronger odour due to initial biodegradation within the container. This may be particularly acute where wet pulverised waste is involved.

5.70 Three principal ways of minimising the generation of landfill odours are:

(a) wastes should not be deposited into standing water,

(b) good compaction and suitable gradients to minimise water ingress together with provision of adequate cover and,

(c) ensuring immediate deposition of the waste on delivery.

5.71 If it becomes necessary to take ameliorative action, options available include:

(a) improving drainage,

(b) spreading hydrated lime over newly filled or saturated waste,

(c) de-odourisers, although their smell may be perceived by some as being worse than that from decomposing waste,

(d) flaring of landfill gas,

(e) increasing cover thickness or using different cover material.

Fires

5.72 Waste materials should not be burnt on landfill sites. Although decomposing putrescible waste may give rise to temperatures up to 60°C within the waste mass, fires on landfill sites are now uncommon. The main ways in which wastes can be ignited are:

(a) burning or hot incoming waste,

(b) discarded cigarettes from smoking on site (see paragraph 5.51),

(c) totters,

(d) spontaneous ignition, and

(e) vandalism.

5.73 Incoming wastes which are found during inspection to be hot or on fire prior to deposition should be directed away from the filling area to a location where the material can be extinguished. Trespassers should be discouraged from gaining entry to the site at all times and the operators should be made aware of the dangers of discarding burning cigarettes or matches (see also Annex 2 paragraph 2.8).

5.74 Surface fires may be extinguished with water, earth or foam. Underground fires are more difficult to deal with. One approach is to dig out the burning refuse and to blanket it with earth. Isolation of the burning area by the excavation of trenches and allowing the waste to burn out may contain the fire. Leachate can be used to fight fires if it is available. The local Fire Authority should be informed when fire occurs on a landfill site and in any case must have been consulted on suitable precautions prior to the site being opened. Emergency procedures to be followed in the event of fire should be displayed at every site.

5.75 Fires are particularly difficult to extinguish at sites receiving baled waste since the fire tends to spread along the interface between bales. It will be necessary to move and isolate the bales; this may be difficult (see paragraph 5.26). Where fires could give rise to hazards such as noxious fumes, contingency plans and emergency procedures are essential.

5.76 As discussed in Chapter 3 (paragraph 3.75 et seq), a risk of fire or explosion arises from the formation of landfill gas. Gas control measures will have been specified in the working plan for the site. Operating practices that should be employed to avoid the risks are described in Annex 2.

Monitoring

5.77 Monitoring is an essential part of landfill operation, not only during the operational life of the site but also throughout the aftercare period. One aspect of monitoring, ie the measurement of waste input, is referred to in paragraph 5.9. Another aspect is the monitoring of the physical behaviour of a landfill. As discussed in Chapter 6, a considerable amount of settlement can take place at landfill sites. Settlement will have been estimated before starting operations, since such estimates are needed to achieve the final contours specified in the planning permission for the site. Settlement should be monitored regularly to ensure that it is even and that it is taking place at the rate predicted; if it is not, remedial measures may need to be taken.

5.78 Monitoring of leachate and landfill gas formation, and environmental monitoring needed to confirm the adequacy of leachate control measures, will have been specified in the working plan. Monitoring requirements associated with leachate management are discussed in paragraph 3.67 et seq; the monitoring of landfill gas is referred to in paragraph 3.84 and is discussed in more detail in Annex 2.

CHAPTER 6: SITE RESTORATION

Introduction

6.1 Disposal by landfill is the major disposal route for many wastes, including controlled wastes and it is important that the completed landfill satisfies both official and public expectations. No matter how well and how scientifically the site has been selected, prepared and operated, it is by the standard of restoration and the continuing satisfactory performance of the restored land that the acceptability of landfill will be judged. Poor restoration can all too easily alienate public confidence in landfill.

6.2 Restoration must be a prime consideration from the outset of landfill site development since the intended ultimate use of the land will influence site design, preparation and operation in various ways. Considerations that arise include: soil stripping and its storage, progressive restoration, final landform including settlement, revegetation and aftercare including leachate and gas management. Some or many of these considerations may feature as conditions in the planning permission or disposal licence and should have been included in the working plan for restoration (see paragraph 3.120).

6.3 The following advice is aimed at all landfill operations, although it is recognised that its relevance will to some extent depend on the stage of restoration already achieved. It must, however, be noted that the advice and guidance given here can be only of a general nature. Restoration is site specific and expert advice must be sought to ensure that proper account is taken of the particular conditions that prevail.

The restoration plan

Introduction

6.4 A plan for restoration of the landfill should form an integral part of the overall planning of the landfill operation; indeed it is part of the working plan of the site which is submitted in support of the planning and disposal licence applications. The plan should influence the conditions attached to the planning permission and the disposal licence. Designing a restoration plan is a complex task which needs to draw on a wide variety of skills to be successful. It is important that the restoration plan meets both the environmental and planning needs of the area, while at the same time paying due attention to costs. A good landfill restoration plan will draw on the expertise of people such as planners, soil scientists, land drainage engineers, hydrogeologists, landscape architects, horticulturalists, construction engineers and specialist farm contractors.

General aspects of the plan

6.5 The content of the restoration plan will depend to some extent on the proposed afteruse, although all plans should include an assessment of the expected settlement and estimates of the amount of extra waste that will be necessary to allow the landfill to settle to the planned final contours. Also, all plans should outline the means to be adopted to deal with leachate and landfill gas. Consequently, the initial restoration plan may require some subsequent detailed development, especially in the light of experience during site preparation. However, in any event it is essential to have established a general restoration plan at the commencement of the landfill operation. This plan, while specifying a use or uses for the restored landfill, need not be too detailed. Some flexibility is essential. During the lifetime of a modern landfill, which is often in excess of 20 years, conditions and requirements will almost certainly change. The restoration plan should set out the framework for restoration and afteruse, while the conditions attached to the planning permission and disposal licence should allow the submission of detailed schemes at appropriate times. The framework should set out the broad type of afteruse intended for the site and how this is to be achieved in general terms. If progressive restoration is adopted, which this report recommends, then the initial plan should specify in some detail the restoration of the first one or two working areas. An example of how restoration might be incorporated into a working plan is given in Figure 6.5.

6.6 Progressive restoration in particular lends itself to subsequent improvement in practices as a result of experience and monitoring. Whenever possible this should be the practice adopted. For example a plan might be specified whereby, in Figure 6.5 phase A is completed and restored, while phase B is being infilled. However, only when phase A has been restored to the required specification is phase B restored and area C made available for infilling. In this way not only is restoration made an integral part of the landfill operation, but the experience gained in restoring part of the site can be incorporated into the detailed restoration schemes prepared later for other areas of the site.

Restoration to vegetation

6.7 Where a proposed afteruse requires some form of revegetation the plan (see paragraph 6.72 et seq) should, where appropriate, also provide an evaluation of the soil quality present at the site. For those landfills where no soil is present, the plan should give an indication of how and where the soils or soil substitutes required will be obtained. The plan should also deal with such matters as the amount of soil needed, soil drainage, vegetation schemes and integration of the restoration into the

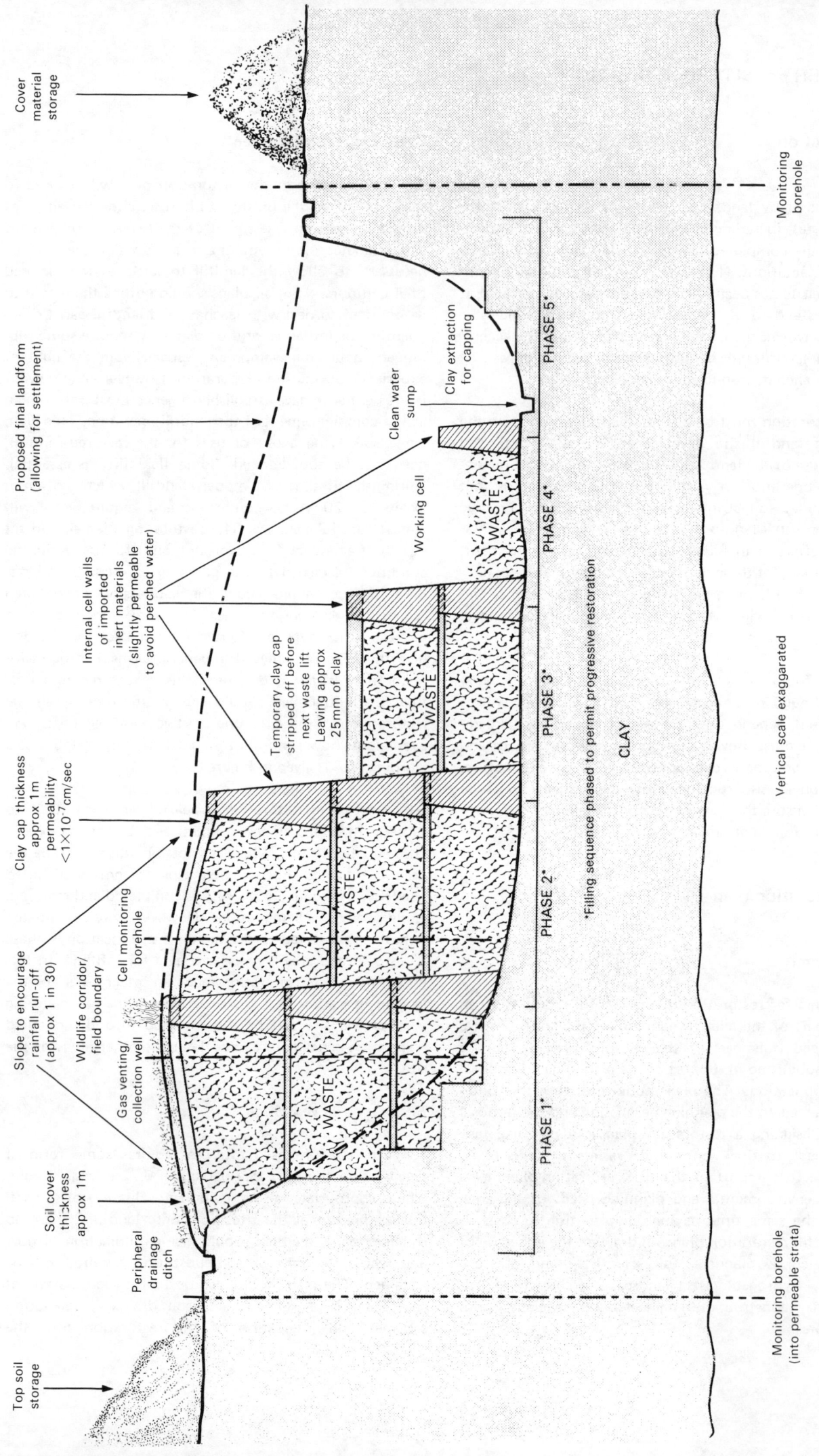

Figure 6.5   SECTION THROUGH LANDFILL SITUATED ON CLAY STRATA

Cover material storage

Monitoring borehole

Proposed final landform (allowing for settlement)

Clean water sump

Clay extraction for capping

PHASE 5*

Working cell

WASTE

PHASE 4*

Internal cell walls of imported inert materials (slightly permeable to avoid perched water)

Temporary clay cap stripped off before next waste lift Leaving approx 25mm of clay

WASTE

PHASE 3*

*Filling sequence phased to permit progressive restoration

CLAY

Vertical scale exaggarated

Clay cap thickness approx 1m permeability <1×10⁻⁷ cm/sec

WASTE

PHASE 2*

Slope to encourage rainfall run-off (approx 1 in 30)

Wildlife corridor/ field boundary

Cell monitoring borehole

Gas venting/ collection well

WASTE

PHASE 1*

Soil cover thickness approx 1m

Peripheral drainage ditch

Monitoring borehole (into permeable strata)

Top soil storage

surrounding landscape. The extent of restoration that can be undertaken at reasonable financial cost may be limited by physical conditions on the site. Such limitations should be recorded and their implications for restoration assessed.

Restoration for construction

6.8 Building on a restored landfill is generally not recommended, at least for a considerable period after its completion, although following assessment building on shallow old landfills may be possible. However, in a few instances the circumstances are such that use of the restored site for building is the originally designated after-use. In such instances, in addition to settlement, leachate, landfill gas evolution and surface and sub-surface drainage the restoration plan will need to include such topics as the design and construction of foundations, provisions to safeguard services and the types and size of buildings that are acceptable. Building developments on landfill sites which have received substantial amounts of difficult wastes is not recommended. (See also paragraph 6.84 et seq).

Post restoration management

6.9 Landfill site restoration cannot be considered to be complete after soils have been placed and the land engineered to the desired contours. For afteruses requiring vegetation there needs to be a recuperation period during which the land is carefully managed and cropped to help the soil recover from the effects of movement, storage and replacement. Indeed the Town and Country Planning (Minerals) Act 1981 empowers mineral planning authorities to specify up to a five year aftercare period for mineral workings being restored to agriculture, forestry or amenity use. Processes operating within a landfill do not cease simply because the site has been restored. A restoration plan should therefore have considered the types of aftercare and monitoring facilities that may be required, who will be responsible for providing them and how they are to be financed.

Financing restoration and aftercare

6.10 Restoration costs for a landfill are likely to lie within a range from £15,000 per hectare to £30,000 per hectare (1984 prices) or some 2 to 5 per cent of total landfill costs (see Annex 4). The actual cost will be determined by such factors as the type of restoration, quantity of soil imported, engineering works to deal with leachate and landfill gas, and the cost of laying, treating and vegetating the restored surface. The cost of restoration will have to be met from the revenue accruing from operation of the landfill site and appropriate provisions made to accumulate funds for this purpose. The restoration plan enables the amount and timing of the required financial provisions to be estimated and facilitates determination of the optimum arrangements. Aftercare of the site will also need to be financed, and this too should feature in the restoration plan. Examples of restoration costs are given in Annex 4.

Landfill processes affecting restoration

Introduction

6.11 A landfill site can be regarded as a complex reactor where physical, chemical and biological processes occur and thereby convert or degrade waste, eventually to provide stablised land. The period needed to achieve biological stabilisation will depend on many factors including the type of waste landfilled, its degree of compaction and moisture present in the waste. Such processes will continue for some years after waste deposition ceases and can be of great significance to successful site restoration.

Settlement

6.12 The intended final (post-settlement) levels and contours of a landfill should be indicated when the restoration plan is drawn up. Contours will normally be specified in planning permission conditions. In determining final land form, a need for it to be restored to its original contours may not necessarily be appropriate, for example, either where raising the final land level is aesthetically desirable or where records of original levels do not exist. It may also be in the best interests of successful restoration to modify contours to allow for adequate drainage or for visual effect. Particular problems may arise in river washland areas where requirements by a Regional Water Authority to avoid flooding problems are likely to preclude raising land levels above original contours, even to allow for subsequent settlement back to these contours. (In Scotland such problems are dealt with by provisions of the planning system.)

6.13 One practical problem faced by a landfill operator is to achieve approved final levels. To do this successfully it is necessary to anticipate the amount of settlement that will occur and to ensure that it occurs as evenly as possible. Even settlement is a crucial factor in successful restoration, indeed if this is not achieved it is unlikely that the restoration will be entirely successful.

6.14 Settlement is influenced by physical, chemical and biochemical factors. It is therefore important to ensure that waste is deposited in such a way that significantly different types of waste do not occur in discrete pockets, as this can lead to differential settlement. Generally the extent of settlement due to physical factors will depend largely on the degree of compaction achieved.

6.15 Aerobic and anaerobic breakdown of biodegradable constituents present in waste also causes settlement. Such processes reduce the particle size of waste, the smaller particles filling any remaining void space, thereby leading to settlement. Bacterial decomposition of waste leads to leachate formation and usually, landfill gas. The escape of landfill gas represents a loss of matter from the landfill. Since household refuse comprises roughly 50% organic materials, of which as much as 40% may be carbon; assuming all the organic matter was easily degradable, and the landfill gas produced was 60% methane, 40% carbon dioxide; landfill gas generation could account for the loss of as much as 22% of the weight of dry refuse. Such a loss would result in very considerable settlement. In practice, however, the weight loss will be less than the theoretical maximum and take place over a considerable period of time eg. up to 20 years. Leachate migration from a landfill, or its abstraction, will also account for losses which are less easy to estimate. However, the rate of landfill gas production and leachate generation are interdependent. The removal or migration of leachate therefore, may affect the rate of landfill gas generation since both landfill gas and the organic gas producing fractions in leachate are derived from waste decomposition.

6.16 The rate and degree of settlement occurring at a landfill will tend to be site specific on the types and quantitites of waste deposited. Sufficient data are therefore not available to enable these factors to be estimated with the accuracy required for planning restoration. Measurements should therefore be made during the lifetime of the landfill and for some period after closure. Experience at a landfill where wastes were compacted to only a moderate density of 0.6 tonne/m^3 showed that settlement could be as much as 35% initially and might average in total 17% of the refuse volume deposited at the end of 10 years. However, at another landfill where a thin layering technique was used, settlement over the first year of measurements was much lower. The results from measurements at a number of landfill sites suggest that total final settlement could lie somewhere between 10 to 25% of the depth of the landfill.

6.17 It is better to overestimate settlement, though the need to achieve the levels of the final planned land surface must always be borne in mind. On the basis of current knowledge, the following observations may assist in predicting settlement. They assume good site management and control, a mix of dry household, commercial and industrial wastes, and the use of compactors:

(a) landfills will normally contain wastes differing in age and deposits near to the base of a landfill may achieve final settlement early in the working life of the site due to surcharges and changes in water content, both of which can influence the rate of degradation;

(b) settlement will mostly have taken place within 10 years, but may not be finally complete for up to 30 years;

(c) wastes compacted to medium densities of some 0.6 tonne/m^3 will settle by up to 20%;

(d) wastes compacted to maximum in-situ densities of 1.2 tonne/m^3 will settle by up to 10%,

(e) each subsequent layer of refuse will help compact the layers below. On the basis of experience in Cambridgeshire, each subsequent 2 metre layer can reduce by 1 year the time taken by the layer(s) below to reach full settlement.

Some worked examples based on these rules of thumb are in Appendix 6a.

Leachate

6.18 Leachate production and treatment has been discussed in Chapter 3. The problems that can arise after restoration should be borne in mind in determining the practices to be adopted during the preparation and operation of a landfill. From the view point of restoration and aftercare, measures to prevent ingress of water both during and after landfilling operations are to be preferred. Systems that rely on the collection of leachate inevitably have long-term implications for aftercare. While leachate strength diminishes with time, very few landfills produce a leachate which, after operations cease, is acceptable for direct discharge to surface waters. Some form of treatment and its associated costs is therefore almost inevitable. Arrangements for providing for such aftercare of restored landfills are not fully provided for in current legislation, and the operator or landowner should agree who is to assume responsibility for this (see Chapter 2, paragraph 2.67).

6.19 If leachate is allowed to rise into restoration soils then severe vegetation loss will be inevitable. On landfills where building construction has subsequently taken place after restoration, damage to services and concrete foundations may occur. In either case the leachate level must be lowered, and action taken to prevent a recurrence. There is no inexpensive way to achieve this.

Gas

6.20 The formation and control of gas produced by landfills is discussed in Chapter 3 and Annex 2. Gas generated within a landfill will seek and normally find a means of escape to the atmosphere. On many landfill sites well isolated from buildings and habitation, it may be practicable although not desirable to allow the gas to vent in a natural random manner. Such a practice if allowed, can result in areas of vegetation damage due to deoxygenation of the soil atmosphere and other effects of the gas. The timing and duration of such occurrences cannot be predicted with any accuracy.

6.21 In all other cases it will be necessary to control where the gas will vent. If the site strata permits lateral diffusion of the gas it may be necessary to prevent its escape laterally from the site by means of a low permeability barrier. Alternatively, the gas may be vented at a perimeter trench backfilled with highly permeable material such as rubble (see paragraphs 3.81 and 3.82). At sites where building construction is allowed to take place it will amost certainly be necessary to direct gases to specific venting points where it may be flared off, collected or simply allowed to vent to atmosphere.

Capping

General

6.22 The uncontrolled release of leachate from a landfill site is one way in which restoration and afteruse might be compromised. A primary aim of restoration is to control and minimise leachate generation by minimising water ingress into the landfill. This can be achieved by installing a low permeability cap over the whole site to increase surface water run-off. Capping is particularly important at landfill sites where difficult wastes have been deposited and those where the subsurface is naturally of low permeability or which have been engineered to contain any leachate generated, for example, by the installation of a liner. Capping may also be used to facilitate landfill gas control or collection. It should be noted that the prevention of water ingress will tend to reduce the degradation rate of organic wastes and hence delay stabilisation of the site.

6.23 Landfill site capping should be constructed of material having a permeability of 1×10^{-7} cm/sec or less. A cap should be domed or contoured to encourage surface water run-off, covered with soil as soon after installation as possible and vegetation cultivated promptly to help prevent water infiltration and encourage water loss by transpiration. To aid the installation of a cap, the final layers of waste deposited may also be domed.

6.24 Clay is most frequently used for capping, but bentonite, a specially treated clay and colliery shale have also been used. Pulverised fuel ash suitably treated to reduce its permeability (by blending with lime or cement) is also a possible alternative. Synthetic materials such as those used for lining landfills have also been used but have the disadvantage that they can easily be damaged by activities on the restored site and unlike clays have no self-sealing properties (see also Appendix 4a).

6.25 The required thickness of a cap will depend on the material used and the in-place permeability that can be achieved. A thickness of about 1 metre for natural materials has been found to be generally effective, practicable and able to cope with most afteruses. The success of capping depends, to some extent, on the way the site was operated. Uneven settlement of the waste can be a major cause of cap failure. Where large amounts of settlement are inevitable it may be advisable to delay placement of a permanent cap. In such circumstances temporary capping should be provided to minimise water ingress.

6.26 To assist in maintaining its integrity a cap should be protected on both its upper and lower surfaces. Accordingly, before a cap is emplaced the surface of deposited waste should be graded and any irregular objects should be removed. In providing a firm base to allow compaction of the cap and to minimise damage from below, a buffer layer should be installed. Where a synthetic material is to be used for capping, a buffer layer at least 0.5 m thick is usually required. Inert material, which does not react with the waste or cap, may be used as a buffer provided that it is free from large stones and lumps. At the same time it should not be so fine that it can permeate into the waste. Coarse or a mixture of coarse and fine gravel may be suitable. Designs for capping should include consideration of leachate and landfill gas collection wells or vents and measures must be taken to ensure that these do not provide a route for water infiltration or gas escape.

6.27 Capping should not be undertaken in very wet weather. The capping material should contain sufficient moisture to ensure good compaction and accordingly spraying with water may sometimes be necessary for materials such as clay or colliery shale. Maximum compaction can be achieved by placing the cap in layers of no more than 0.3 m thickness, followed by rolling. To ensure good integration with the succeeding layers, spraying with water may be necessary. The permeability of the completed cap should be measured to ensure that it meets the design criteria.

Protecting the cap

6.28 For the cap to remain effective it must be protected from agricultural machinery, drying and cracking, plant root penetration, burrowing animals and erosion. An appropriate thickness of soil or soil-like cover material is therefore necessary.

6.29 The required depth of cover over a cap will depend on the intended agricultural afteruse. Approximate depths of soil affected by a range of agricultural activities are:

Shallow sub-soiling relief	400 mm
Deep sub-soiling	450–600 mm
Mole drainage	400–600 mm
Piped field drains	600–1000 mm
Main drains/ditches	1000–1500 mm

6.30 Allowing a cap to become dry and crack compromises its integrity. Small scale cracking in clays occurs soon after drying begins. Plant roots are capable of drying clays to below the moisture content at which cracking occurs. While most plant roots are within the top 300 mm of soil, they are nevertheless capable of drying soils to a depth of approximately 700 mm. On this basis the depth of soil cover used to protect a cap from this effect should be at least 1 m thick. Compacted clays have properties different from natural soils particularly with regard to the ability of roots to penetrate them but precise data on this are lacking. Rooting depth can be controlled by using

Figure 6.39 NAMING SOIL HORIZONS

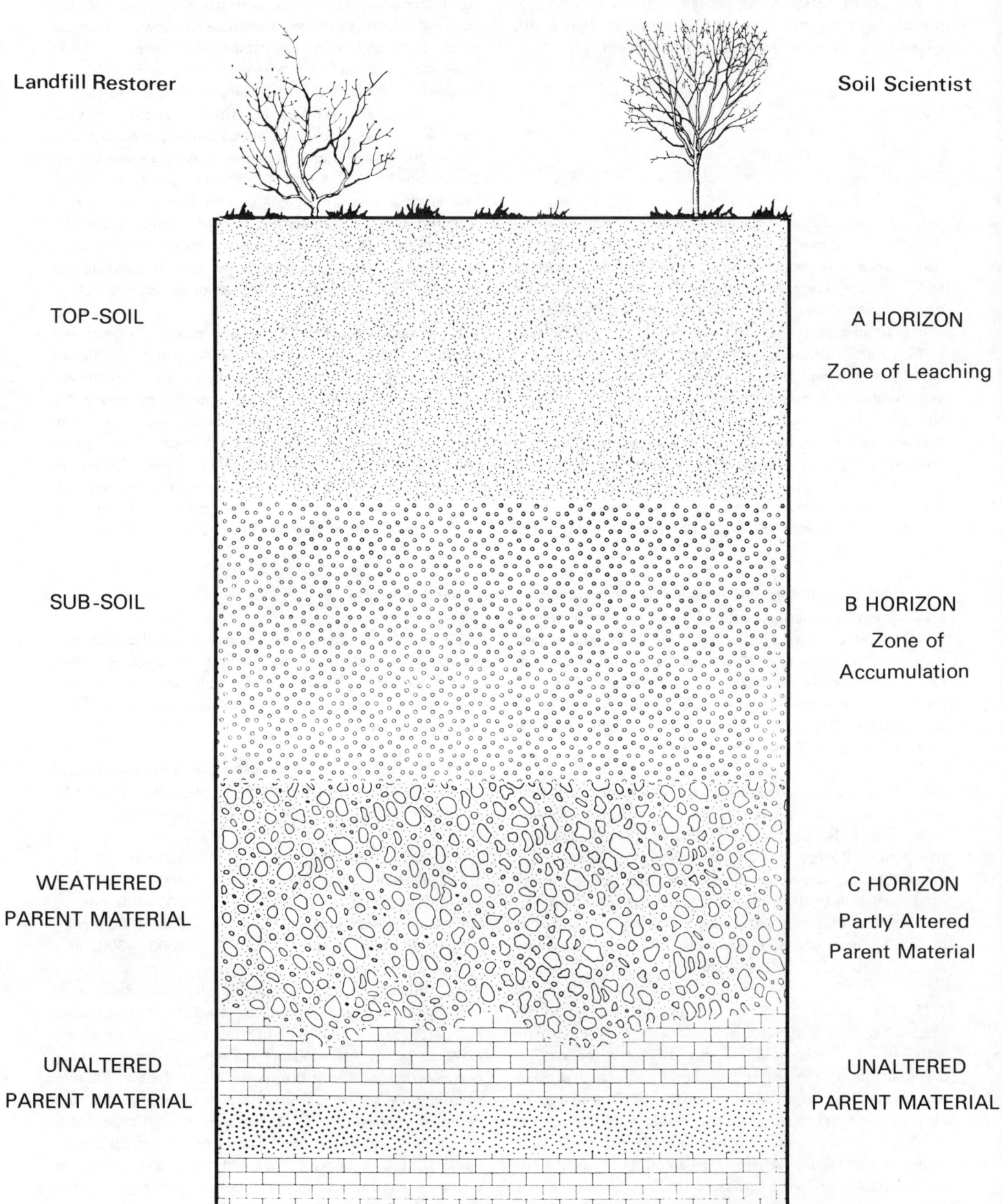

Landfill Restorer

Soil Scientist

TOP-SOIL

A HORIZON

Zone of Leaching

SUB-SOIL

B HORIZON

Zone of
Accumulation

WEATHERED
PARENT MATERIAL

C HORIZON

Partly Altered
Parent Material

UNALTERED
PARENT MATERIAL

UNALTERED
PARENT MATERIAL

species having shallow rooting systems, maintaining low fertility at depth and compacting the base layers of the final soil cover below the depth required to support the vegetation. Vegetation management can also affect soil drying.

6.31 Plant roots will rarely penetrate into light textured soils having a bulk density of greater than 1.7 to 1.8 g/cc, or heavy textured soils with a bulk density greater than 1.5 to 1.6. In order to achieve a permeability of at least 1×10^{-7} cm/sec for a clay cap, its bulk density is likely to be approximately 1.8 to 1.9. Thus, there is a reduced likelihood of root penetration of the cap providing it is protected by sufficient cover material to remain intact and ensure that it does not crack.

6.32 Some trees, especially when self sown, can have roots extending to depths of 5 m or more but even these will have difficulty penetrating a well compacted intact clay cap. However, trees growing in the relatively shallow soil cover above a cap, lack such deeply penetrating and anchoring tap roots. This renders them susceptible to wind throw and may result in a significant disturbance of the cover soil and possibly exposure of the cap. For this reason the planting of trees in soils above clay caps is not recommended.

6.33 The animals which are most likely to cause damage to the integrity of caps are rabbits and the like, which can burrow through cover layers. Earthworms can also create small channels in the cap through which water may percolate. While little investigation has been carried out on the effects of such activities, it is considered unlikely that these will be significant at sites with at least 1 m of soil cover.

6.34 Soil cover will also protect a cap from erosion and the effects of weathering. The cover itself will resist erosion if it supports well established vegetation.

Drainage of capped sites

6.35 Percolation by rainfall through soil cover will cause the water content of the soil above a low permeability cap to rise and eventually exceed its field capacity. The excess water will flow down-slope through the soil cover, either to be collected by a drainage system or, if soil permeability or slope is not sufficient, to saturate the soil and form perched water tables or even ponding and flooding. Factors affecting drainage include soil texture, compaction, slope and intended afteruse.

Capillary movement of water

6.36 Capping is effective in breaking the hydraulic continuity between water contained in the deposited waste and that in the surface soil. Upward capillary movement and escape of leachate is therefore inhibited. Toxic species present in leachate cannot therefore concentrate in the soil cover. Accordingly, the protection afforded by cap-

ping is important at sites where liners have been installed or where difficult wastes have been deposited.

Venting of landfill gas and vapours

6.37 A low permeability cap will impede the upward movement of landfill gas and vapours and should be designed and provided in conjunction with any gas control or collection system. The buffer layer below a cap may allow lateral gas movement below the cap and enhance gas collection. Should there be a need to compensate for a reduction in moisture ingress effected by the cap perforated gas collection pipes may be used for the controlled irrigation of the waste to stimulate landfill gas formation.

Soils

Introduction

6.38 As noted above (paragraph 6.28) a final covering of soil or soil-like material is necessary to protect the cap. Where the projected afteruse requires cultivation of vegetation another consideration arises, that is, the quality of the soil or other material as a growing medium in which plant roots will live and grow.

6.39 Soil is characteristically differentiated into horizons or layers of different appearance and properties. For the purposes of landfill restoration, three major horizons need to be distinguished:

(a) Topsoil — the biologically active surface layer which is cultivated and to which lime and fertilisers are added.

(b) Subsoil — the less well structured and less biologically active layer, but which acts as a reserve of nutrients and moisture.

(c) Weathered Parent Material — the material from which soil is formed.

An idealised diagram showing these horizons and relating them to the nomenclature used by soil scientists is given in Figure 6.39.

6.40 The on-site availability of soil for restoration purposes will vary according to whether the landfill is:

(a) in an old mineral excavation, or similar site such as a disused railway cutting, where little or no soil remains; or

(b) to take place on previously undisturbed land or as part of the restoration of a mineral working where soil has been retained.

Obtaining good quality topsoil and subsoil may prove both time-consuming and expensive. There is, however, some evidence to suggest that some subsoils, if well managed, can provide a medium for adequate plant growth when demanding agricultural crops are not to be grown. Very occassionally some afteruses such as certain types of amenity restoration may not require topsoil to be provided.

6.41 Careful husbanding of resources can greatly reduce the need to import large quantities of suitable soil when landfilling is complete. In the past, topsoil and subsoil have been buried at the bottom of the excavation or used for intermediate cover, even when they have been in short supply. Soils delivered to a site during its operational life, have tended to be used at the time, rather than better quality materials being stockpiled for the restoration phase. However, it must be recognised that space constraints at some sites may prevent storage of such materials.

Soil characteristics

6.42 Soils consist of sand, silt and clay particles and organic matter. The organic matter derives from decaying plant and animal material and hence is most abundant in the topsoil (3 to 8%). Subsoil has less organic matter (around 1 to 2%) and the weathered parent material none. The mineral fraction of soil can be subdivided on the basis of particle size range as shown in Table 6.42.

6.43 The fertility of soil is a function of its type, structure, organic matter content and nutrient levels. A good soil structure having a continuous network of pores and fissures is also essential to allow drainage of water, free movement of air and unrestricted development of plant roots. When soil is wet it can easily be compacted. Frequently this results in the soil surface becoming waterlogged during periods of wet weather, even though under drainage may be present, because of the loss of pore space. Compaction can also restrict root penetration so that a smaller volume of soil is exploited. Crops growing on compacted soils are therefore more susceptible to shortages of water and nutrients and tend to give reduced

yields particularly in a dry season. Soil conditioners may be useful in some circumstances.

Soil stripping

6.44 When a site to be used for landfilling possesses normal soil profiles, these should be removed and stored for use during the restoration. Topsoil, subsoil and weathered parent material should be stripped, handled and stored separately, using appropriate methods, such that they can be replaced in the correct sequence on completing restoration. Such activities should, ideally, be undertaken when soil and weather conditions are suitable.

6.45 Scrapers are commonly used for soil stripping. It is essential that the depth of cut of the scraper is carefully controlled to follow closely any variations in the thickness of soil layers over the site. When stripping soil, the aim is to minimise any damage to soil structure which could be caused by smearing and compaction. Smearing results from the slipping or spinning of wheels or tracks, by pressure from vehicles forcing soil aggregates to slide past each other or by the movement of a stripping blade across the soil. Soil compaction is caused mainly by the ground pressure of vehicles traversing it. Generally, tracked vehicles apply the lowest ground pressures and hence the least compaction, while large self-propelled earthscrapers particularly when changing direction are probably the most damaging. Soil removed by backactor and dump-truck usually results in minimal soil compaction. Since most compaction occurs during the first passage of a vehicle over uncompacted soil, should repeated movement of vehicles be necessary it is better that repeated traverses use the same track marks. At all times vehicles should ideally run on the lowest soil profile to be exposed.

6.46 Both smearing and compaction are related to soil moisture content. To minimise soil structure damage, stripping operations should take place only when the soil is in a reasonably dry condition. Stripping operations should cease both during and for a period after heavy rainfall. Criteria for the control of soil movement need to be determined for each site individually since soils differ so much in character. Subsoils are often less resistant to

Table 6.42 Soil particle size

	Diameter	Benefits for restorations involving vegetation	Disadvantages to restorations involving vegetation
Stones	2 mm +	May possibly help drainage	Dilutes the volume of genuine soil. Interferes with cultivation.
Sand	0.06 to 2.0 mm	Increases permeability Good moisture retention if sand is fine	Chemically inert. Poor moisture retention if coarse.
Silt	0.002 to 0.06 mm	Good moisture retention	Unstable soil structure.
Clay	0.002 mm	Chemically very active	Sticky when wet Very hard when dry.

After McRae, 1979.

compaction because of their lower organic contents. Generally, most topsoils remain sufficiently dry to permit stripping operations for much of the summer. However, difficulties may be experienced during prolonged dry conditions to achieve satisfactory movement of subsoil.

6.47 Damage to soil structure mostly occurs during stripping operations as the soil is picked up and off loaded. An obvious way to reduce this is to minimise handling. Direct movement of soil from the area being stripped to another area being finally restored is therefore preferable to double handling into and out of storage. This type of direct operation is normally possible only when a landfill is being progressively restored.

Soil storage

6.48 Soil is normally essential to the restoration phase of a landfill site. Consequently, both preparatory to and during landfilling operations, stockpiles of soil need to be accumulated. Initially, one way this may be achieved is to construct screening banks. During operations suitable soil received for disposal can be segregated and stockpiled. In some instances it may be necessary to import and stockpile suitable soil. When constructing soil stockpiles excessive compaction can be reduced by creating low wide mounds rather than tall and narrow ones. It is likely, however, that soils will deteriorate during storage by becoming anaerobic. It has been observed that during prolonged storage noticeable deterioration can occur at depths below 0.3 m from the surface of stockpiles of clay like soils and below about 2 m depth in sandy textured soils. Associated chemical changes affect soil pH, the available nutrients and organic matter concentrations. Bacteriological activity in stored soils diminishes with depth and earthworms will be absent. If soils are to be stored for long periods, a grass crop will help to minimise surface erosion and maintain a soil structure in the surface layers. Weeds should be controlled with approved herbicides. As a general rule, topsoil mounds should be cultivated and particular care taken to ensure that they are properly drained. Should soils become anaerobic during storage, recent experience indicates that aerobic conditions may be quickly re-established subject to them being properly handled during respreading, and their subsequent good management.

6.49 The creation of soil storage mounds and their locations, together with controls for their management, should have been decided before site operations commenced and controlled subsequently by appropriate conditions attached to the planning consent. Table 6.49 indicates the soil properties which are important for both plant growth and land restoration. It also indicates the ways in which planning conditions may be used to minimise soil damage caused by its disturbance on landfill sites.

Imported soils

6.50 At a large number of landfill sites, particularly those in old mineral workings or on other disturbed or derelict land, there is unlikely to be any original soil left. At other sites additional soil may be required. To obtain this therefore, the operator may either purchase it or alternatively husband soil from construction waste soil materials delivered to the site. The first option is likely to prove extremely expensive and, since good quality soils are always in demand, is not always available. The second option is most frequently adopted. However, it may be necessary to accept delivery of a wide range of construction wastes to ensure that adequate quantities of suitable soil like materials are obtained. Consequently the material being brought in as "soil" is likely to contain considerable amounts of concrete, brick rubble, timber, fencing and other demolition rubbish. The site operator must therefore segregate the soil from this material possibly using mobile separation plant and husband it in separate stockpiles. It is likely to take a considerable time to accumulate the amounts of soil required for restoration. Its accumulation should be started, therefore, soon after the site begins operations. Neglecting to stockpile soils until the site is nearly full or, indeed, is full is not acceptable since it can considerably delay final restoration.

6.51 Some imported soils will be of poorer quality than the indigenous soils in the surrounding area. It may sometimes be possible to improve soil quality by incorporating bulky organic materials such as conditioned or digested sewage sludge, milk whey, brewery wastes or spent hops, spent mushroom compost, or in some parts of the country sugar beet wastes. Many types of organic wastes do, however, require careful evaluation and analysis to avoid the introduction of excessive amounts of undesirable contaminants. There are DOE/National Water Council, MAFF and Scottish Agricultural College guidelines on acceptable concentrations of heavy metals in sewage sludge for application to agricultural land. Analytical details on their sludge can be obtained from the appropriate water authority. Unless carefully managed a number of organic wastes may also give rise to unacceptable smell problems.

Alternative, soil-like materials

6.52 It is possible that at some sites the only readily available material that can be used for restoration, either as subsoil or, in extreme cases topsoil, would not normally be regarded as a soil. For example, quarry wastes, tailings or material from silt ponds, general spoil from excavation, dredgings, fragmentiser waste, RDF rejects and pulverised fuel ash (which is a controlled waste) have all been successfully used at some time for site restoration. When investigating the possible use of such materials, important considerations will include the anticipated afteruse, the likelihood of phytotoxic compounds being present, particle size distribution including the number of stones and other obstructive materials, and the permeability and moisture retention characteristics. Such considerations may therefore rule out a large number of possible sources of such alternative materials. Great care is needed in the choice, placement and aftercare of such materials if a successful restoration is to be achieved.

Table 6.49 Soil characteristics and effects of disturbance

Soil characteristics	Effects of disturbance	Effects controllable by planning conditions
Soil profile and depth: Arrangement and thickness of different horizons (topsoil, subsoil and weathered parent material)	Possible mixing of different soil horizons, loss of material, possible bulking during soil movement and subsequent resettlement	Careful separation, stripping, storage and respreading of soil horizons commensurate with amounts of soil actually present
Soil texture: Size range of primary particles present (sand, silt, clay, etc)	Not necessarily altered if soil movement carefully controlled	Careful separation, stripping, storage and respreading of soil layers
Soil structure: Arrangement of individual soil particles into larger, compound units or "peds" with channels between	Inevitable disturbance by soil movement; extent depending on type of structure and conditions of movement. Compaction; increase in bulk density; impeded drainage	Method of soil movement; avoidance of movement in wet conditions; direct respreading where possible, subsoiling and other cultivations on replaced soil; remedial cropping
Bulk density: The weight of soil per unit volume. A measure of compaction	Possible loosening during stripping decreases bulk density but main danger is increased bulk density by passage of earthmoving machinery over it	As soil structure.
Soil drainage: Movement of water through the soil. Depends mainly on soil texture and structure; and level of water table	Disturbed by soil movement	See soil texture and structure. Levels and gradients of reinstated sites; subsequent installation of drainage system
Available water capacity: Measure of moisture which plants can extract from the soil. Related to soil texture and structure	May be altered by soil movement	Not directly
Nutrient status and chemical characteristics: Content of main plant nutrients (nigroten, phosphorus, potassium, calcium, magnesium), acidity (pH), and micronutrients (such as manganese, copper, molybdenum, iron).	Soluble compounds leached during storage of soils, and pH may be lowered. Anaerobism in wet/compacted soils	Addition of lime and fertilisers, as indicated by ADAS analysis, on replacement of soils and during aftercare period. Occasionally may need fertiliser and lime added to soil stockpiles

Final landform

6.53 The final landform specified in the planning permission and restoration plan will determine the mode of operation of the landfill and the means of achieving successful restoration. The restoration should aim to blend in with the adjacent land yet be shaped to facilitate drainage.

Run-off and drainage

6.54 Run-off depends on many factors including the quantity and intensity of rainfall, the porosity of topsoil and subsoil, the presence or absence of an impermeable layer or hard pan, the presence or absence of vegetation, the type of vegetation and the gradient of the soil profiles. The gradient of the final profile is of particular significance to the drainage of restored land. Artificial perched water tables will be produced above any impermeable layer in the soil. Surface ponding of water may occur when the porosity of the soil is limited or when rainfall intensity exceeds the soil permeability. Such drainage problems can best be overcome by creating soil profile

gradients, in the form of a dome. By doing so, water will be shed laterally to the edges of the site where if necessary a system of boundary ditches or drains can be provided. An estimate of likely quantitites of run-off water from severe storms needs to be calculated to provide a drainage system with sufficient capacity. It is suggested that the run-off likely from at least a 1 in 10 year storm be estimated to arrive at the required capacity of a drainage system to serve the landfill. Removal of water can also be affected by the installation of piped underdrainage systems. These are more easily installed and are more effective if a gradient has already been provided for in the restoration designs. Separate planning permission may be required to allow suitable gradients to be achieved.

6.55 Opinions vary on the gradients needed. In general, final settled gradients in the region of 1 in 30 have been found to be satisfactory to prevent ponding and drainage problems created by differential settlement. Less steep slopes may be acceptable on sites that have taken only inert wastes. Steep gradients should be avoided wherever possible. Land to be used for arable crops should have slopes of less than 1 in 6. Slopes of 1 in 3.5 tend to be the limit for the use of agricultural machinery. Steep slopes may increase the risks of erosion, especially on land where the vegetation cover has not become established.

6.56 The afteruse of a restored landfill site will dictate the drainage requirements necessary. In some instances it may not be necessary to install land drains especially for those landfills where unimpeded infiltration of rainfall into the waste is acceptable. In such instances it is advisable to avoid over compaction of the final waste layer and the soil used for restoration, thereby avoiding the need for installed drainage.

6.57 Artificial drainage schemes may be necessary in some circumstances. When required, such drainage systems should be installed preferably during the final restoration phase. In the past, the laying of drains has sometimes been delayed until subsidence has ceased. Nowadays, however, with better compaction and the use of plastic rather than tile drains, provided that a sufficient gradient can be achieved, there appears to be no reason to delay drain laying. However, it must be borne in mind that piped drainage systems need to be covered by a sufficient depth of soil, (at least 600 mm), to avoid damage by agricultural machinery and problems of freezing in the winter.

Soil replacement

6.58 The placement of soil to allow cultivation subsequently is probably the most critical stage in site restoration. Soils must be emplaced to minimise damage to the soil structure. The restoration plan should specify the quality of soil to be used, where they are to be obtained and, the order and depths to which they are to be emplaced. In determining soil thicknesses it should be borne in mind that disturbed soil frequently possesses a lower bulk density than undisturbed soil.

6.59 Minimal compaction may be achieved by handling and spreading soil materials when they are in a suitable physical state to do so. Dry soils are inherently strong, increased moisture reduces their bearing strength and consequently when wet soils are subjected to pressure they are easily compacted and can form a puddled mass. The remedial measures needed to repair such damage can be significant.

6.60 Even when soils are replaced under favourable conditions, some compaction will inevitably occur particularly when conventional earth scrapers are used. The following procedures have been found to be effective in minimising such compaction. The placement of soil should start at the furthest point from the site entrance ensuring that the scraper hauls over the surface of the fill and not over subsoil and topsoil layers already emplaced. Reinstatement of the soil should be carried out in strips. Compaction created by earth movers is not only the obvious compaction caused by the passage of the rear wheels but also includes compaction of the previously

emplaced soil layer by the front wheels (Figure 6.60). Its compaction will be hidden by the newly laid material. Effective control over the movement of site traffic is essential.

6.61 The most effective disturbance of a previously compacted soil can be achieved using a winged subsoiler. However, if its tines are set too deep the soil can fail to shatter and may merely deform instead. The success of subsoiling depends on the moisture content of the soil. Subsoiling wet soil is likely to leave slots in which moisture will remain below the surface while subsoiling very dry and hard soil will produce large clods of soil. During subsoiling operations care must be taken to ensure that the integrity of any impervious cap is not destroyed.

6.62 Compaction can be avoided almost entirely if soils are moved and emplaced using hydraulic excavators (backacters) and dump trucks working from the surface of the fill. The surface of the fill should have been cleared of debris. Soil should be laid in strips, the width of which will be determined by the size and type of machinery available.

6.63 Soil should be loaded from the stockpile into dump trucks, using a tracked hydraulic excavator or wheeled or tracked loader. These dump trucks should transport the soil to strip being restored and tip it in heaps on the surface of the fill. From a position on the surface of the fill the soil should then be spread and levelled using a tracked hydraulic excavator with a wide bucket. When the first strip of subsoil has been levelled, topsoil should be brought in and placed in heaps on the fill adjacent to the levelled subsoil. From here it can be lifted by the excavator and spread over the subsoil to the required depth. When the topsoil has been laid over the first strip, the second strip is started using the same cycle of operations. When a sufficiently large area has been topsoiled, a lightweight bulldozer or tracked loader should be used to grade the restored surface.

6.64 Scrapers, though not entirely suitable from a soil care point of view, probably offer the quickest method of removing, stockpiling and replacing soils. If weather and soil conditions are favourable, they can also be considered satisfactory from a soil care point of view. However, it is important that the machine operator understands the objectives of the restoration, otherwise he may sacrifice soil care for speed of operation.

Returning landfills to vegetation

Introduction

6.65 Experience has shown that landfills can be successfully returned to some form of vegetation, whether for agriculture, forestry or amenity and such afteruses are expected to remain the most common specification for landfill restoration in the foreseeable future.

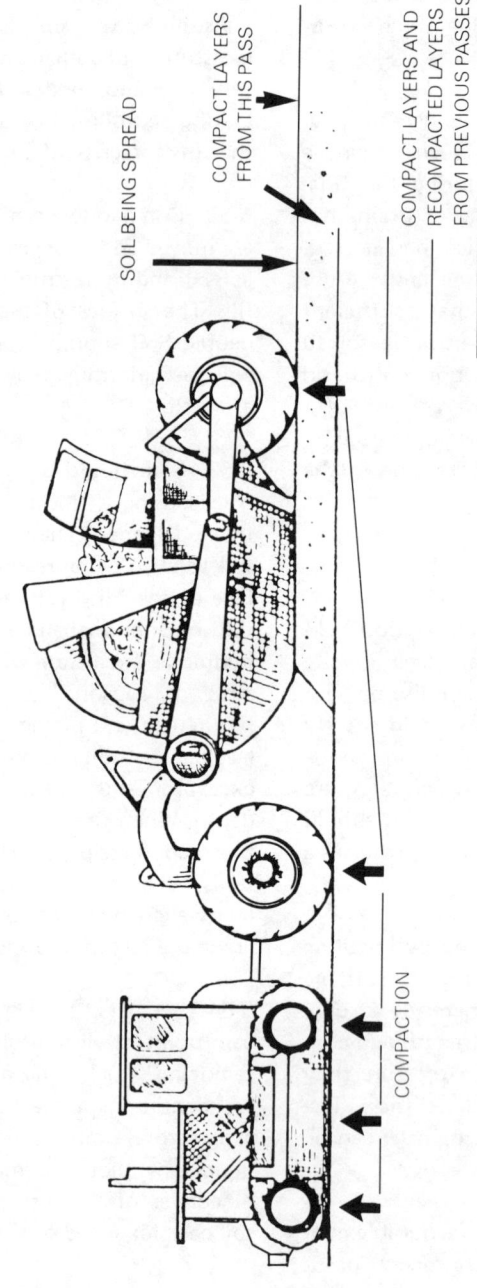

**Figure 6.60   COMPACTION OF SOIL WHILST SPREADING
After McRae (1979)**

SOIL BEING SPREAD

COMPACT LAYERS
FROM THIS PASS

COMPACT LAYERS AND
RECOMPACTED LAYERS
FROM PREVIOUS PASSES

COMPACTION

These compact layers need to be disrupted mechanically by pulling a tined implement through the soil
i.e. a 'subsoiler' or 'ripper'

Soil specification

6.66 Where agriculture is the planned afteruse, the aim should be for the restoration to match, as far as possible, the quality and productive potential of the surrounding land. In the rare situation where better quality land has to be used, the Ministry or Department of Agriculture concerned will recommend strict requirements for restoration soil depths and particularly the thickness of topsoil to ensure that aims are achieved. If the restored land is to fall within either Grade 1 or 2 of the Agricultural Land Classification (Appendix 6b), such requirements could prove extremely difficult to meet. A considerable depth of high quality soil, such as loams, with the capacity to retain optimum moisture whilst at the same time being capable of freely draining excess water. Such soils would have to overlay a sufficient thickness of subsoil which itself would need to possess good drainage and moisture retention properties. Piped land drains would probably be necessary. It is for such reasons that Grade 1 or 2 land should be used for landfill only where there is no other alternative.

6.67 Most arable crops ideally require a 1 metre depth of soil to provide sufficient moisture reserves for optimum growth; Table 6.67 gives rooting depths for some of the more common plants used in restoration. As noted in paragraph 6.30, a depth of at least 1 metre of soil is needed over a clay cap to avoid it being damaged by drying out and cracking. A substantial depth of soil is also required to protect a membrane or clay cap against damage by subsequent agricultural operations (see paragraph 6.29). Where protection of a cap is not involved, the soil depth may be determined largely by crop yield considerations. It is possible to grow crops on a minimum depth of 500 mm of settled soil. However, piped underdrainage will not be possible within such a depth of soil without restricting cultivation. For more demanding crops such as sugar beet a minimum depth of 750 mm of soil may be essential. Alternatively for some forms of restoration, container grown plants can be used. When these are planted, with an adequate root ball of soil, the thickness of surrounding soil may be decreased. Such a planting scheme will inevitably be expensive.

Table 6.67 Rooting depth of plants commonly grown on restored landfills

Plant	Rooting depth
Cereals	up to 1 m.
Potatoes	600 to 700 mm.
Sugar beet	over 1 m.
Lucerne	over 1 m.
Grasses	usually up to 300 mm but in some species can be considerably greater.
Trees	mainly within the first metre but in some self sown species the tap root can extend several metres.

6.68 If the afteruse of the site is for grassland or recreation purposes a lesser depth of soil can be tolerated but plant growth may be reduced under drought conditions. On land intended for recreational purposes, sharp objects can work their way up through a thin soil layer and could be dangerous or a nuisance to users. Following levelling and covering or capping, a minimum depth of 100 mm of settled topsoil is required for playing fields although a greater thickness (up to 300 mm) is preferable. For football pitches in particular, a relatively free-draining sandy soil is desirable. If piped land drains are required, a minimum depth of 600 mm of settled soil above the drains will be necessary.

6.69 The possible loss of soil by erosion must be considered and due allowance made when specifying soil depths. Erosion by wind can occur on exposed sites, particularly when they are not cultivated for long periods of time. Sites restored to steep slopes can be subject to serious erosion by the effects of rainfall. The use of soil bonding agents should be considered in such situations (see also paragraphs 6.34 and 6.54).

Soil improvement

6.70 About 15 different mineral nutrients are essential for plant growth. Some are required in relatively large quantitites (macro-nutrients), while others, the trace elements (micro-nutrients), are required only in minute amounts. The main macro-nutrients are nitrogen, phosphorous, potassium, calcium, magnesium and sulphur. The more important trace elements are iron, manganese, copper, zinc, boron and molybdenum.

6.71 Soil acidity is also an important factor in plant growth. Most natural soils have pH values between 4 and 8, the optimum for most arable crops being pH 6.5 and for grassland pH 6.0 although values between pH 5.0 and 5.5 are amenable to some kinds of turf. Low pH values can result in reduced yields, while high pH values reduce the availability of trace elements which also can reduce crop yields. The addition of lime to soil raises its pH value and hence is used extensively for this purpose on acidic soils. Another important property of lime is its beneficial effect on soil structure, since it induces the flocculation of soil particles (particularly clays) to form a friable structure. Fertiliser also need to be applied and advice on their use may be obtained from fertiliser suppliers.

Planting options

6.72 The main planting options for restored landfill sites are:

(a) grass for grazing,

(b) grass for fodder crops (hay and silage),

(c) grass for sports or amenity use,

(d) trees for amenity use or forestry,

(e) arable crops,

(f) assisted natural revegetation for amenity use.

Grass

6.73 The mixture of grass seed applied will vary according to the afteruse. It is doubtful if there is a "best" mixture for all landfills. However, in establishing grass for grazing or fodder, the seed compositions given in Appendix 6c have been found to be generally suitable for landfill site restoration. Regardless of the grass mixture used, it has been found advantageous to apply seed at much higher rates than those normally used in agriculture. Seeding rates of around 45 kg/ha (40 lb/acre) may be necessary.

6.74 In choosing seed mixtures for amenity or sports ground use it is advisable to seek advice either from the Sports Turf Research Institute or seed companies. The selection of suitable seed mixture depends upon the use to which the turf is to be put, taking into account the position of the site and its landfill history, particularly in regard to the wastes deposited and any problems met with leachate and gas. Some of the mixtures which have proved successful in landfill site restoration for amenity use are given in Appendix 6d.

6.75 Reinstatement to some form of grassland is the restoration option most likely to be successful. The quality of soil and its degree of preparation is generally less demanding than other afteruses since grass tends to be more tolerant towards adverse slopes, drainage and the like than other species. The root structure of a grass sward also helps to stabilise and improve the soil structure. The dung and urine from grazing animals are also beneficial and there is no need for below-surface cultivation. Particular care must however be taken to avoid hoof damage (poaching) by grazing animals.

Trees

6.76 Conditions attached to many planning permissions require that a specified programme of tree planting is undertaken. Trees can enhance amenity areas and also provide a habitat for wildlife. They can also be grown as a long term commercial investment. However, such a requirement cannot be generally recommended, particularly at landfill sites which have received difficult wastes.

6.77 While trees can be grown satisfactorily, by virtue of the depth of cover used being sufficient to provide enough moisture storage during periods of prolonged dry weather, waste materials deposited on landfill sites, either because of their chemical properties or physical nature, can provide an inhospitable environment to their roots. Almost inevitably, roots will penetrate into deposited waste at some

stage and are unlikely to survive. For similar reasons and to maintain deposited wastes in isolation, trees should not be planted or allowed to grow on lined landfill sites or where water input is regulated by an impervious cap or membrane. Where trees are grown around the perimeter of such sites, they should be positioned at a sufficient distance from the landfill to ensure that impermeable barriers are not compromised.

6.78 Oxygen is essential to tree-root growth. For trees to remain healthy 18% or more of oxygen in the soil voids is desirable; less than 12% will lead to tree death. The presence of landfill gas in soil is therefore significant. Its main constituent, methane, is not toxic to tree roots, but its presence will reduce the concentration of oxygen in the soil atmosphere. By contrast carbon dioxide, the other main constituent of landfill gas, is toxic to roots, and its presence in concentrations greater than 5% to 10% in soil atmospheres is likely to lead to tree death. Also, the high temperatures, occasionally in excess of 40°C, that are found at landfills when degradation is proceeding are not conducive to plant root growth and survival.

6.79 Soil pH should always be considered when selecting trees for planting at a particular location. Appendix 6e shows suggested species of trees and shrubs suitable for a range of soil pH values. Most trees will grow satisfactorily on a range of soils without the need to add fertilisers but if the pH is less than 4.5 liming may be necessary to extend the range of species which can be grown. The application of phosphatic fertiliser prior to planting may be advantageous. Better survival rates are obtained when small trees are planted. Weed control is essential and trees may need staking until they become established.

Arable crops

6.80 Land used for the cultivation of arable crops requires better initial soil quality and care than the afteruses so far discussed. The depth of soil required is principally to allow agricultural machinery to be used and to provide sufficient moisture for growth. Table 6.67 gives rooting depths for some of the more common plants and crops used in restoration. The requirement of a good quality soil takes account of the inability of most arable crops to contribute greatly to the improvement of soil structure and stability.

6.81 The method of cultivation employed together with the seeding rate influence the degree of success achieved in cultivating arable crops on completed landfill sites. The results from a study suggests that the use of full cultivation techniques may be more beneficial on restored soils than the minimal cultivation techniques common to current agricultural practice. Minimal cultivation employs only one pass with a chisel plough and with discs.

6.82 For restoration either to grassland or arable crops, good weed control is important, particularly during the first few years. Weeds can be expected to compete successfully with plants and crops and unless held in check can easily overcome the best re-vegetation and restoration schemes. A wide range of chemical weedkillers and herbicides is available and advice on suitable types and their rate of application should be obtained from the suppliers.

Natural revegetation for amenity

6.83 In some urban situations, acceptable restoration has been achieved by planned encouragement of the site to revegetate naturally. Such an approach should not be considered to be an abandonment of high standards of landfill practice. By encouraging and husbanding natural revegetation the aim is to create an area of countryside for the benefit of those living within the urban zone. However, this form of restoration needs a long time-scale and may need to be protected from unwanted outside influences until it becomes established.

Building on landfills

Introduction

6.84 Landfills may contain voids where efficient compaction has not been achieved. More importantly, such landfills will be subject to differential settlement. Landfill leachate and gas also may adversely affect structures and buildings foundations constructed on them. Recently completed landfill sites will present greater construction problems than old landfill sites, mainly as a result of the longer time period for putrescible waste to have decomposed, settled and stabilised at old sites and the fact that changes in the composition of household wastes have occurred over recent years. Since the 1930s the ash content of household refuse has decreased appreciably while the paper, rag, metal, plastic, glass and putrescible content has increased.

6.85 Scarcity of land in urban areas and pressure to bring back derelict land into beneficial use is leading to some landfill sites being considered for building development. While the difficulties of building on recently completed modern landfills may be extremely expensive to overcome, shallow landfills completed 20 or more years ago may, subject to detailed assessment, be found to be suitable. Building has taken place on old landfills, though the number of such developments and the short time that has elapsed since construction, makes it difficult to judge its success.

6.86 Engineered containment landfills should not be built on. Any development is likely to destroy the integrity of the impervious cap and piling through the landfill to the underlying strata inevitably will compromise the containment. Such landfills are also likely to contain a high liquid content. This has particular implications with respect to the provision of services (particularly deep foul sewers), the stability of structures and attack on construction materials.

Site investigation

6.87 Before deciding whether or not a landfill can be built upon it is necessary to assess its suitability, including the underlying strata, for the proposed development. In this way potential problems can be identified at an early stage and an evaluation made of the likely performance of the deposited waste as a foundation material. The assessment should identify problem areas including not only geotechnical factors but also the possibility of chemical attack on foundations, gas and leachate generation, combustibility and toxicity. All information relating to waste disposal previously undertaken at the site should be reviewed followed by ground investigations, including trial pits and boreholes and vertical and lateral stability. Trial pits reveal the nature of the fill, its composition and likely variability while boreholes can provide valuable information on water levels in the refuse and gas emission rates. Investigation should also be conducted to determine the rate of settlement. Simple field loading tests may be appropriate.

Nature of fill

6.88 One important consideration requiring investigation is the amount of biodegradable matter present in the landfill. If the percentage of such material is high there is likely to be considerable problems associated with building on the site. The homogeneity of the waste material is also important. Other significant properties include chemical composition of the waste and the presence of combustible material.

Degree of compaction

6.89 Compaction is largely a function of the method of landfill operation used; even well compacted refuse will have a lower density than that of soils and needs to be evaluated during site investigations for development.

Landfill depth

6.90 Generally the greater the depth of waste landfilled the more severe will be the problems that are likely to be identified. Sudden changes in the depth of fill, for example if the site was originally terraced, can cause serious problems of differential settlement.

Landfill age

6.91 Recently deposited refuse is likely to provide a poor foundation material. Not only has there been a shorter time for decay and decomposition processes to have taken place but, as noted above (paragraph 6.87), decomposing refuse has inherently poor geotechnical characteristics.

Settlement

6.92 Building on landfills presents several geotechnical problems. Where refuse has been placed above natural ground level the stability of the slopes at the edge of the landfill needs to be considered. As stated in paragraph 6.89 above the bearing capacity of the fill should also be investigated. However, the major geotechnical problem encountered is likely to relate to long-term settlement. A basic distinction needs to be made between settlement of the refuse caused by the loads applied by buildings constructed on it and settlement due to degradation processes operating within the landfill. In general the latter type of settlement will predominate and the concept of bearing capacity directly related to settlement caused by applied loads may be misleading. Foundation designs should be based on estimates of differential settlement. It must be noted that at a number of sites baled refuse has been deposited. The long-term behaviour of such sites is not yet known and the figures for settlement and other factors which influence the decision on whether or not to allow construction may not apply to such sites.

Other problems and hazards

6.93 Problems can arise with landfills containing combustible material both from the fire risk and from the generation of methane. Methane is a combustible gas and an explosion hazard can exist where for example, gases accumulate in a confined space such as under the floor slab of a building. In this situation the two basic approaches to gas control are to provide ventilation and the installation of impermeable barriers. Ventilation methods used include perforated pipes driven into the waste and trenches filled with hard core around the site. Because of the hazards associated with landfill gas in buildings, developments on sites producing landfill gas should be permitted only in exceptional circumstances. Only designs which will ensure the safety of the buildings and their occupants until gas is no longer being produced — possibly 50 years — should be used. Structures should be designed to provide gas-tight membranes beneath the ground-floor slab, taking due account of the difficulty of ensuring a truly gas-tight seal where services and drains penetrate the floor slab. It is usually more practicable for services to enter a building above ground level rather than through the floor slab. On landfills where methane emission rates are significant it may be necessary to locate the floor slab above ground level and arrange for natural or forced ventilation beneath it. The provision of gas detection equipment and the regular monitoring of methane concentrations both under floor slabs and within buildings erected on landfill sites is essential.

6.94 A fire hazard can exist by virtue of the presence of uncompacted combustible waste materials, the voids in which can provide sufficient air to support combustion underground. Since the supply of air is likely to be small, the rate of combustion inevitably will be slow. Sources of ignition can include the deposit of burning waste materials, the deposit of pyrophoric materials eg finely divided metal turnings, fires lit on the surface of a site, or self-heating and ignition. Oxidising agents which may be present in some wastes could provide sufficient oxygen to initiate spontaneous combustion. It is important that any boreholes provided in landfilled waste are properly grouted up and the grouting checked periodically particularly if settlement occurs.

6.95 While little is known about the durability of materials used in the construction of buildings on landfill sites, the landfill environment could be a particularly aggressive one for those materials that are commonly used. Leachate may attack such materials; the usually high sulphate content of leachate is particularly important because of its adverse effect on cement and concrete structures. Codisposed wastes (see paragraph 7.7) may affect cement or concrete, though by how much is again difficult to predict. Variations in pH between 5.5 and 8.5 do not seem to pose any threat to concrete and cement. Indeed most good quality concretes seem able to withstand a pH of down to 3.5 and up to 9, though a pH higher than this will adversely affect some concrete aggregates. A flow of leachate past concrete will increase its susceptibility to acid attack. Guidance on the behaviour of concrete in sulphate-bearing soils and groundwaters is provided in Building Research Establishment Digest 250.

6.96 Exposed metals used in building construction are also likely to be attacked, and they should therefore be coated with resistant materials. Certain plastics and other organic-based materials may be degraded in a landfill environment.

Foundation design

6.97 A number of possible approaches to the problems previously described can be suggested. As a drastic solution, deposited waste might be excavated, removed from the site and replaced by an inert fill imported from elsewhere. The new fill should be emplaced in thin layers and well-compacted.

6.98 Another way of overcoming settlement problems is to use piled foundations and a suspended floor. The piles should be designed to allow for downdrag caused by settlement of the refuse. In landfills where methane is being generated, piles could provide pathways for the escape of gas. Special attention is needed in the design of services which enter buildings founded on piles, so that the floor slab is not penetrated. The material used to construct the piles also needs to be considered.

6.99 Ground improvement techniques available are essentially methods of increasing the density of the fill. Such methods include pre-loading with a surcharge of fill and the specialist techniques of "dynamic consolidation" and "vibroflotation" are designed to increase density and make the fill less compressible. Such improvement methods are, however, likely to be suitable only on waste in which decay and decomposition is largely complete and which is not saturated with leachate.

6.100 One system of dynamic consolidation typically involves repeatedly dropping a 15 tonne weight from heights of up to 20 metres. Primary tamping usually consists of repeated impacts at a number of points on a widely spaced grid. In later stages a more uniform tamping is carried out over the whole area with a reduced drop height.

6.101 "Vibroflotation" uses a cylindrical poker which contains in its bottom section an eccentric weight. Rotation of the weight results in vibrations in a horizontal plane being transmitted to the fill and the poker penetrates into the waste fill. The long cylindrical hole produced by the poker vibrator is backfilled with granular material. The method has thus two effects. It compacts the waste and reinforces it with columns of well-compacted granular material. However, where there is a large amount of putrescible material present, the effectiveness of this method can be questionable since the columns derive their strength from the lateral restraint provided by the surrounding soil. In decomposing refuse this may be inadequate.

Post-restoration management

The need for post-restoration management

6.102 Restoration cannot be regarded as complete when soils have been replaced and vegetation planted. Several years of management and treatment may well be required. Care must be taken to husband the restored site until it has fully recovered from the impacts of the restoration itself. Consequently, post-restoration management needs to be carefully planned and implemented. Continued management is necessary at former landfill sites in order to remedy differential settlement, to provide for post closure monitoring and if required, the treatment and disposal of leachate. At many sites it will also be necessary to monitor and control the emission of landfill gas. Where the planned afteruse is for agriculture, forestry or amenity use a management scheme should be prepared specifying the steps to be taken to bring the land to the required standard and maintain it for such use.

Legal provision for post-restoration management

6.103 Building on completed landfills is discussed at 6.84 et seq where it is generally concluded that it is not an advisable practice. However, it must be recognised that in time a restored landfill can blend in with the environment to such an extent that on the surface it will be indistinguishable from its surrounding landscape. In such a situation proposals for development may be made in ignorance of the fact that the land had previously been used for waste disposal and degradation processes may still be taking place. As such it is recommended that records are kept to provide details on all land used for waste disposal purposes.

6.104 Powers are available under planning legislation to attach conditions to the planning permission to cover the restoration and afteruse of the landfill. Additional provisions contained in the Town and Country Planning (Minerals) Act 1981 may be applied where planning permission for mineral working is also a permission for restoration landfill waste disposal. A mineral planning authority may apply an aftercare condition or scheme which requires up to five years management for land restored to agriculture, forestry or amenity use. The aftercare that may be specified under the Act may include planting, cultivating, fertilising, watering, draining or otherwise treating the land. Advice on aftercare, including the statutory consultations required, is given in DOE Circular 1/82.

6.105 While a disposal licence is in force powers exist under the Control of Pollution Act 1974 for action to be taken to remedy any possible emergency that might arise on a waste disposal site through inadequate aftercare. Once a disposal licence has been surrendered or revoked, the powers under which the landowner can be required to put his site into a satisfactory condition include the following:

— the Health and Safety at Work etc Act 1974 in terms of immediate security;

— the enforcement of any relevant planning conditions;

— an action under section 79 of the Public Health Act 1936 to require the removal of accumulations of noxious matter (for example, leachate);

— an action under sections 92 to 100 of the 1936 Act where a statutory nuisance is involved;

— an action in tort under common law where a private nuisance is alleged; or

— an action under section 16 of the Control of Pollution Act 1974 where waste has been deposited in breach of licensing conditions.

6.106 Current legislative provisions for the restoration and aftercare of landfills are recognised as being inadequate in certain aspects and are being reconsidered.

Restoration of landfills that have taken difficult wastes

6.107 The principles underlying the restoration of landfill sites that have accepted difficult wastes are the same as for sites which have received only household or inert wastes, therefore, the basic considerations presented in this chapter will apply. However, the presence of difficult wastes raises additional considerations to ensure that potentially hazardous or toxic materials are not released in an uncontrolled manner. These considerations are presented in Chapter 7.

ESTIMATING SETTLEMENT (see paragraph 6.12 et seq and 6.92)

The following example, though simplified indicates how planned layering can lead to site restoration to an acceptable level. However, there are too many variables to allow them to be used other than as a guide.

Consider a site based on the following assumptions:

(1) original site depth — 9.00 m

(2) waste placed in 2 m deep layers

(3) annual settlement rates: (Note: the following rates are purely conjectural and in reality will depend on many factors such as; the types of waste and their proportions and consistency on delivery, methods and thickness of emplacement etc.

Years after deposit	Annual settlement rate %	
1	10.0	— when 2.00 m settles to 1.80 m
2	6.0	— when 1.80 m settles to 1.69 m
3	4.0	— when 1.69 m settles to 1.62 m
4	3.0	— when 1.62 m settles to 1.57 m
5	2.0	— when 1.57 m settles to 1.54 m

(4) total final settlement — 20%

(5) restoration:

— clay cap thickness 1.00 m
— top and subsoils thickness 1.00 m

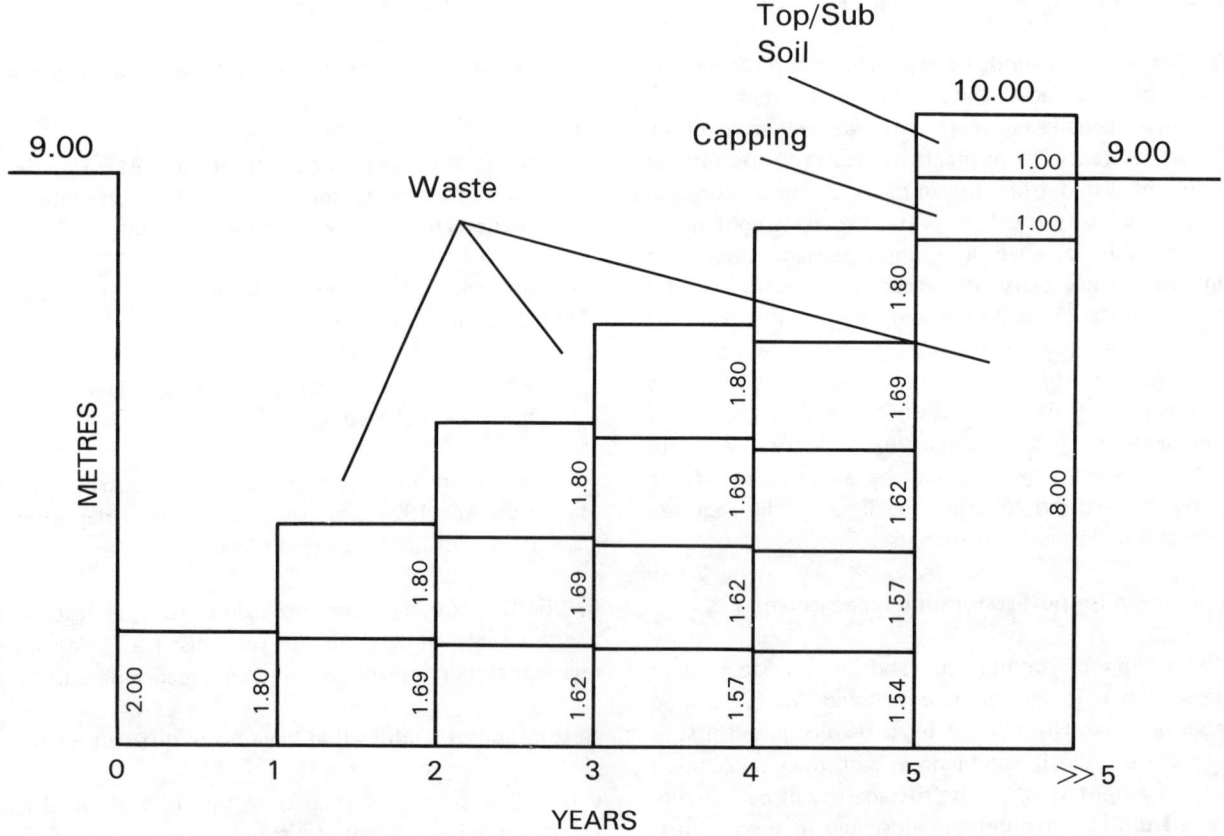

Depth of waste after 5 years = 8.22 m plus 1.0 m capping and 1.0 m restoration soils, giving a completed level of 10.22 m which ultimately settles to 10.00 m. The final site contours would be achieved by adjusting the thickness of the final waste layer across the site to allow eg. doming, drainage slopes, continuity of final landfill surface with surrounding land surfaces etc.

THE AGRICULTURAL LAND CLASSIFICATION OF ENGLAND AND WALES

Grade 1

Land with very minor or no physical limitations to agricultural use. The soils are deep, well drained loams, sand loams, silt loams or peat, lying on level sites or gentle slopes and are easily cultivated. They retain good reserves of available water, either because of storage properties of the soil or because of the presence of a water table within reach of roots and are either well supplied with plant nutrients or are highly responsive to fertilisers. No climatic factors restrict their agricultural use to any major extent. Yields are high on these soils and cropping flexible since most crops can be grown, including the more exacting horticultural crops.

Grade 2

Land with some minor limitations which exclude it from Grade 1. Such limitations are frequently connected with the soil; for example its texture, depth or drainage, through minor climatic or site restrictions such as exposure or slide may also cause land to be included in this grade. These limitations may hinder cultivations or harvesting of crops thus leading to lower yields or making the land less flexible than Grade 1. However, a wide range of crops can be grown, though there may be restrictions on the range of horticultural and arable root crops on some types of land in this grade.

Grade 3

Land of average quality, with limitations due to the soil, relief or climate, or some combination of these factors which restrict the choice of crops, timing of cultivations or level of yields. Soil defects may be of structure, texture, drainage, depth, stoniness or water holding. Other defects such as altitude, slope or rainfall may also be limiting factors. The range of cropping is comparatively restricted on land of this grade. Only the less demanding horticultural crops can be grown and towards the bottom of the grade, arable root crops are limited to forage crops. Grass and cereals are thus the principal crops and the land is capable of growing reasonable yields when judiciously managed and fertilised.

Grade 3a

Moderate degree of limitation common to Grade 3, but has some physical advantages which lead to appreciably better performance than that of land in the remainder of the grade.

Grade 3b

Most of this land is capable of average production.

Grade 3c

This land has physical characteristics which give a poorer performance than that of other land in this grade.

Grade 4

Land with severe limitations due to adverse soil, relief or climate, or a combination of these. Adverse soil characteristics include unsuitable texture and structure, wetness, shallow depth, stoniness or low water holding capacity. Relief and climatic restrictions may include steep slopes, short growing season, high rainfall or exposure.

Grade 5

Land of little agricultural value with severe limitations, due to adverse soil, relief or climate, or a combination of these. The main limitations include very steep slopes, excessive rainfall and exposure, poor to very poor drainage, shallow depth of soil, excessive stoniness, low water holding capacity and severe plant nutrient deficiencies or toxicities.

SEED COMPOSITIONS FOUND EFFECTIVE WHEN PLANTING RESTORED LANDFILLS TO GRASS FOR GRAZING OR FODDER (see paragraph 6.73)

	Percent by weight
Mixture A	
Italian rye grass	30
Perennial rye grass (medium early) (*Lolium perenne* spp *gremie*)	25
Perennial rye grass (medium late) (*Lolium perenne* spp *combi*)	25
Timothy (*Phleum pratense*)	20
Mixture B	
Perennial rye grass (medium early) (*Lolium perenne* spp *gremie*)	25
Perennial rye grass (medium late) (*Lolium perenne* spp *combi*)	25
Perennial rye grass (late) (*Lolium perenne* spp *melle*)	25
Red Fescue (*Festuca rubra* spp *I. Horalis*)	10
Timothy (*Phleum pratense*)	10
White Clover (*Trifoluim repens* spp)	5
Mixture C	
Perennial rye grass (medium early) (*Lolium perenne* spp *gremie*)	18
Perennial rye grass (medium late) (*Lolium perenne* spp *combi*)	18
Perennial rye grass (late) (*Lolium perenne* spp *melle*)	20
Timothy (*Phleum pratense*)	15
Cocksfoot (*Dactylis glomerata*)	15
White Clover (*Trifolium repens* spp)	14

Notes

1. White Clover — small leaf for highest persistency. Medium leaf variety has high resistance to disease.

2. Red Fescue has been used extensively on hill land reclamation.

3. Cocksfoot is drought resistant and may be valuable on fast draining land in areas where rainfall is low.

SEED MIXTURES FOR AMENITY RECLAMATION (see paragraph 6.74)

1. The choice of seed mixtures will be influenced by the site and soil conditions. For amenity afteruses, the following components will be important:

(i) the inclusion of white clover on all sites except possibly those with good quality topsoil,

(ii) the use of perennial ryegrass on sites where hard wear is anticipated; but using only low-maintenance cultivars in most cases (see note 1),

(iii) the use of bent/fescue mixtures where maintenance is to be infrequent,

(iv) the avoidance of perennial ryegrass and other coarse grasses in mixtures where wild flowers are to be encouraged (note 3),

(v) the use of ecologically adapted species if the final soil cover is particularly acid (below pH 5.0) or alkaline (above pH 7.0).

2. Optimum seed application rates will vary considerably, depending on soil type, aspect, slope and the required final use. They will generally be higher than the normal agricultural rates, and fall within the range 50-200 kg/ha.

3. The following seed mixtures and sowing rates are examples for three different, general amenity afteruses for neutral soils (pH 5.5-7.0). They are based on a Countryside Commission publication, which contains more detailed information (see note 2) and with subsequent advice and updating from the Sports Turf Research Institute.

(i) General informal recreation areas

Such areas may cover a wide range of uses but in general a hard wearing sward of medium height is required.

	Percent by weight
Smooth stalked meadow-grass (*Poa pratensis*)	30
Perennial ryegrass[1] (*Lolium perenne* cultivars)[1]	25
Timothy (*Phleum pratense*)	15
Slender creeping red fescue (*Festuca rubra* ssp *Litoralis* etc)	10
Chewings fescue (*Festuca rubra* ssp *commutata*)	10
White clover (*Trifolium repens*)	10

Sow at 60-80 kg/ha

(ii) areas of visual importance

These areas will normally be used only lightly for recreation; and the aim will be to provide medium to long grass which may be cut between one and three times a year.

	Percent by weight
Slender creeping red fescue (*Festuca rubra* ssp *litoralis* etc)	40
Chewings fescue (*Festuca rubra* ssp *commutata*)	30
Common bent (*Agrostis tenuis; Agrostis castellana*)	20
White clover and common bird's-foot-trefoil, plus wild flower seed[3] (*Trifolium repens, Lotus corniculatus,* and required wild flower mix)[3]	10

Sow at 70-100 kg/ha.

(iii) Picnic areas

Require a short, fine-leaved but hard wearing sward with little or no maintenance.

	Percent by weight
Slender creeping red fescue (*Festuca rubra* ssp *litoralis* etc)	25
Chewings fescue (*Festuca rubra* ssp *commutata*)	20
Smooth stalked meadow grass (*Poa pratensis*)	20

Perennial ryegrass[1] or rough meadow grass   15
(*Lolium perenne* cultivars[1] or *Poa trivialis*)

Common bent   10
(*Agrostis tenuis; Agrostis castellana*)

White clover   10
(*Trifolium repens*)

Sow at 60-80 kg/ha.

Notes

1. General advice on grass cultivars, including low maintenance cultivars, can be obtained from the Sports Turf Research Institute (STRI) and major seed houses. The latest advice is contained in the leaflet "Turfgrass Seed 1984", available from STRI, Bingley, West Yorkshire.

2. Countryside Commission (1980). Advisory Series No 13, Grassland Establishment in Countryside Recreation Areas.

3. Wells T, Bell S, and Frost A (1981) "Creating Attractive Grasslands Using Natural Plan Species". Nature Conservancy Council, Interpretative Branch, Attingham Park, Shrewsbury.

TREES AND SHRUBS FOR RESTORATION AROUND LANDFILL SITES (see paragraph 6.76 et seq)

Species selection on their soil pH preference.

1. Trees

(a) pH 6.5–8.0 – Ash, Birch, Cherry, Corsican pine, Field maple, Grey poplar, Hazel, Italian alder, Lawson cypress, Leyland cypress, Norway maple, Red Cedar, Sycamore, Yew, Willow.

(b) pH 5.0–6.5 – Ash, Aspen, Birch, Cherry, Common alder, Corsican pine, Field maple, Grey alder, Grey poplar, Hazel, Larches, Lawson cypress, Locust, Scots pine.

(c) pH 3.6–5.0 – Birch, Common alder, Corsican pine, Grey alder, Larch, Lawson cypress, Locust, Scots pine.

2. Shrubs

(d) pH 6.5–8.0 – Buddleia, Field maple, Guelder rose, Hazel, Privet, Sallows, Thorns, Wayfaring tree, Whitebeam.

(e) pH 5.0–6.5 – Bramble, Field maple, Hazel, Holly, Privet, Sallows, Thorns.

(f) pH 3.6–5.00 – Bramble, Buckthorns, Holly, Sallows, Strawberry tree.

Notes

1. Where cover materials are of poor quality, species such as alders or locust, which can fix their own nitrogen, should be mixed into stands in a suitable ratio.

2. Tree lupin is not at present recommended because of devastating attacks by a newly arrived aphis which frequently kills the plant.

3. Gorse and Broom are not recommended because of the high fire risk when they become old.

4. Field maple can be regarded both as a tree and a shrub and has been exceptionally successful in many roadside plantings.

5. Much useful information on the establishment and aftercare of native tree and shrub species can be found in "Planting native trees and shrubs" by K & G Beckett, published by Harold Colour Publications (1979).

Trees and shrub species for sites where the substrata is aerobic for at least 0.5 m

Pedunculate Oak	*Quercus robur*
Sessile Oak	*Q. petraea*
Silver Birch	*Betula pendula*
Hairy Birch	*B. pubescens*
Black Poplar (1)	*Populas nigra var. betulifolia*
Hawthorn	*Crataegus monogyna*
Blackthorn	*Prunus spinosa*
Willow and Sallow (2)	
Hazel	*Corylus avellana*
Alder	*Alnus glutinosa*
(particularly in wetter locations)	

Notes

1. Black poplar is one of the only two native species of poplar. The other is Aspen (Populus Tremula).

2. As far as is possible, only native willows and sallows should be used. These are, trees: crack willow (*Salix fragilis*), white willow (*S. alba*) and shrubs: goat willow (*S. caprea*), grey willow (*S. atrocinerea* and *S. cinerea*) and purple willow (*S. purpurea*).

Trees and shrub species for sites where the substrata is aerobic for less than 0.5 m

Silver Birch	
Hairy Birch	
Alder	
Sallow	(*Salix caprea*)
Lupin	

Species to be encouraged (and planted) at all sites

Bramble	*Rubus fruticosus*
Ivy	*Hedera helix*
Honeysuckle	*Lonicera periclymenum*

(These species are probably not commercially available but can be easily propagated.)

Further tree and shrub species that can be planted on fertile sites

Purging Buckthorn	*Rhamnus catharcticus*
Alder Buckthorn	*Frangula alnus*
Ash	*Fraxinus excelsior*
Beech	*Fagus sylvatica*
Cherries	*Prunus padus & avium*

Guelder Rose	*Viburnum opulus*
Holly	*Ilex aquifolium*
Field Maple	*Acer capestre*
Wild Rose	*Rosa canina & arvensis*
Whitebeam	*Sorbus aria*

There are other native species that could be planted but are not listed above because of their possible invasive qualities (eg dogwood) or fire-risk (eg broom and gorse). Broom and gorse can be useful species and could be planted if the fire-risk was seen not to be a problem.

As far as possible, the native species that are chosen should be those that are already present in the landscape. The natural ranges of many species does not extend to the whole country and this should also be taken into account.

Much useful information on the establishment and after-care of native tree and shrub species can be found in "Planting native trees and shrubs" by K & G Beckett, published by Harold Colour Publications (1979).

Plates

Plate 1: Landfill site entrance; secure gates and fencing, notice boards, traffic signs, reception cabin.

(Durham County Council)

Plate 2: Fencing; inner – peripheral litter fence, outer – security fence.

(Humberside County Council)

Plate 3: Cell operation; prepared cells prior to infilling.

(ARC Ltd.)

Plate 4: Deep quarry landfill site, leachate collection well in the foreground

(Leicestershire County Council)

Plate 5: Layering and compacting household waste with purpose built compactor.   (AERE, Harwell)

Plate 6: Liquid waste disposal trenches.   (Cleanaway Ltd.)

Plate 7: Site evaluation trials for landfill gas abstraction.   (AERE Harwell)

Plate 8: Landfill lining; 2.5 mm HDPE, leachate collection drains, lined leachate collection pond   (Greater London Council)
     (foreground).

Plate 9: The application of daily/intermediate cover.

Plate 10: Restored Landfill.

(Cumbria Land Reclamation Ltd.)

Plate 11: Landfill gas migration control plant. (Suffolk County Council)

Plate 12: Leachate collection and treatment plant. (Aspinwall and Partners Ltd./Montgomery District Council)

CHAPTER 7: THE DISPOSAL OF DIFFICULT WASTES

Introduction

7.1 This chapter deals with the disposal to landfill of 'difficult wastes'. The term 'difficult wastes' has no statutory basis; it covers a range of industrial wastes the disposal of which calls for special procedures, either because of their hazardous nature or physical characteristics. A list of wastes that can be regarded as difficult, classified according to the system established by the DOE Waste Management Paper No 4, is given in Appendix 7c.

7.2 The principles and practices set out in previous chapters relating to the assessment and design of landfill sites, and to their preparation, operation and restoration, are generally applicable to sites taking difficult wastes. However, as would be expected, such wastes necessitate special consideration at all stages of landfill development. The characteristics of difficult wastes are wide ranging, therefore each site requires to be judged on its merits as to the types of waste it can receive and that appropriate control procedures are adopted; generally, more stringent measures are called for compared to those employed for the disposal of household or similar wastes. Experience indicates, however, that with good management and operating practice, the landfilling of difficult wastes need pose no greater risks to human health or the environment than that presented by, say, household waste.

7.3 This chapter provides guidance on the general procedures that should be adopted for the disposal of difficult wastes but does not attempt to describe in detail the procedures appropriate to the disposal of specific types of wastes. Many wastes and groups of wastes have been considered in detail in the DOE Waste Management Paper series, and include codes of practice for disposal, including where appropriate landfill. In considering specific waste types, reference should be made to the appropriate Waste Management Paper (see Appendix 7a).

Selection of disposal technique

Types of landfill site

7.4 Two broad types of landfill are described in Chapter 3. These are: 'attenuate and disperse' sites (see paragraph 3.8) in which reliance is placed on leachate attentuation processes acting within the landfill, followed by dispersion in surrounding geological strata as any leachate migrates away from the landfill and 'containment' sites (see paragraph 3.10) where the aim is to isolate any leachate from the environment, thus relying on attenuation within the landfill and on leachate management practices to prevent water pollution. Attenuation processes are described in paragraph 3.27 et seq.

7.5 These two basic approaches to landfill site design in principle can apply equally to the disposal of difficult wastes. However, because of the nature of some of these wastes doubts may exist about the attenuating mechanisms that are likely to operate, and hence about the characteristics of the leachate. This in turn will dictate that additional care is needed when considering the use of attenuate and disperse sites for the disposal of difficult wastes. If components of particular wastes attenuate only very slowly or in ways which are not understood a containment approach may need to be adopted, particularly if water resources are potentially at risk. It must be understood that not all wastes will undergo attenuation, no matter how long the retention time. Particular aspects of attenuation applicable to the disposal of difficult wastes are discussed later (see paragraphs 7.15-7.17).

Disposal options

7.6 In addition to having some choice in the type of disposal site to be operated an operator has the choice of four main techniques. These are:

   (a) codisposal,

   (b) monodisposal,

   (c) multidisposal, and

   (d) pretreatment followed by disposal as in a, b, c.

These techniques are described in the following paragraphs.

Codisposal

7.7 Codisposal is the conscious deposition of difficult wastes with household or other similar wastes to achieve specified objectives, a key element of which is to take full advantage of attenuation processes inherent in the landfill site. Central to this objective is the maintenance of a balanced input of wastes to ensure that the attenuation processes are not overwhelmed. Controlling the rate of input of difficult wastes is therefore always necessary. Much of the recent research effort on landfill processes has been concentrated on the effects of codisposing difficult wastes with household refuse. The results obtained have indicated that, when properly managed, codisposal can be regarded as a safe and efficient disposal option for many difficult wastes. (See Figure 7.7).

7.8 Codisposal normally utilises properties available in household-type waste to attenuate those constituents in difficult wastes which are polluting and potentially hazardous and thereby make their impact on the environment acceptable. The practice requires special precautions and management of all operations to ensure that it is both

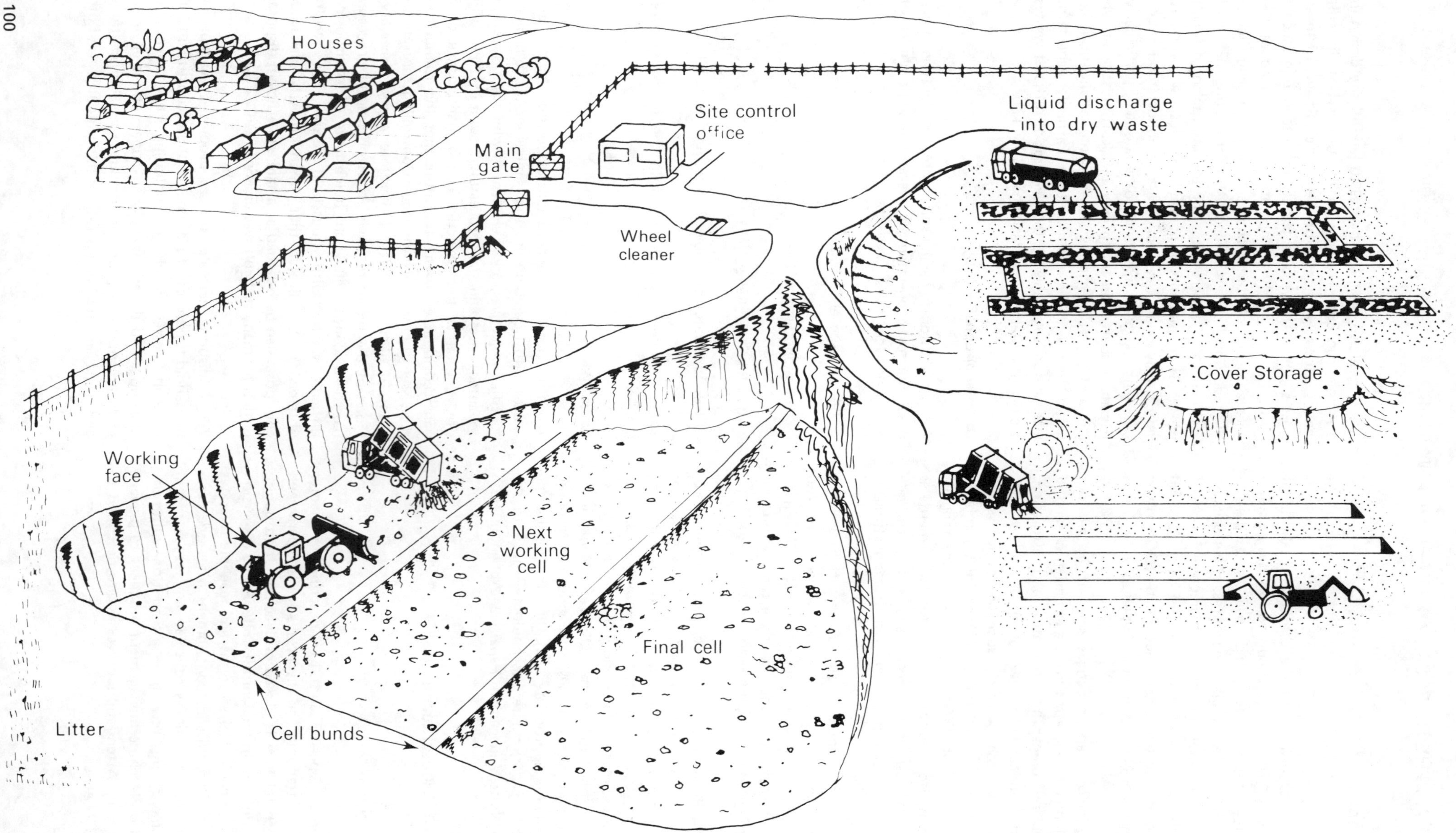

Figure 7.7   TYPICAL CODISPOSAL LANDFILL SITE

Liquid discharge into dry waste

Cover Storage

Site control office

Wheel cleaner

Main gate

Houses

Final cell

Next working cell

Cell bunds

Working face

Litter

safe and environmentally acceptable. Difficult wastes destined for codisposal must be critically assessed and only those wastes which are compatible with household waste should be accepted for codisposal. (see paragraphs 7.11-7.14).

Monodisposal

7.9 Monodisposal is the disposal of wastes having the same general physical or chemical form by landfill or lagooning. Following disposal the waste need not, necessarily, remain in the same physical form as it was produced. For example, Pulverised Fuel Ash (PFA) from power stations is almost always landfilled at monodisposal sites and frequently it is pumped there as a slurry and allowed to dewater. Producers of bulk inorganic chemicals often dispose of large quantities of waste at monodisposal landfills.

Multidisposal

7.10 The aim of multidisposal is to turn deposited mixed heterogeneous waste into a no more hazardous and preferably more environmentally acceptable form than the individual component wastes. Multidisposal is generally used to describe the disposal of chemically different wastes which, like liquids or sludges, have similar physical forms. The deposit of mixed wastes either as liquids into lagoons or at sites accepting both inert and degradable industrial and commercial solid wastes may also be regarded as examples of multidisposal operations.

7.11 Where mixed wastes are to be deposited, it is vital that they should not react to produce either worse polluting or hazardous compounds, or significant concentrations of dangerous or noxious gases and vapours. Detailed chemical knowledge of the wastes and an evaluation of the reaction productions likely to be produced should be an integral part of any multidisposal operation.

7.12 A further type of multidisposal operation is the disposal of difficult wastes into inert industrial solids. While this is similar to codisposal, in this instance attenuation processes within the landfill do not occur.

7.13 Finally, multidisposal has occasionally been used to describe disposal sites which can accept a range of different wastes which are landfilled in separate discrete areas of the site. Mixing of the wastes therefore does not take place. Such a practice is not significantly different from monodisposal other than possibly in scale. Accordingly this type of operation should better be regarded as monodisposal. The term codeposit is sometimes used.

Pretreatment and disposal

7.14 Some wastes are not suitable for landfill disposal in the form in which they arise by virtue of their physical or chemical properties and thereby present an unacceptable risk to personnel or the environment. Pretreatment,

often undertaken before the waste arrives at the landfill, can include simple bulk reduction such as: sludge dewatering to reduce the total quantity of waste, precautionary measures such as the secure bagging of loose asbestos to preclude fibre release during deposit at the landfill, various chemical treatment processes or the use of solidification processes.

Potential impact on water quality

7.15 An appreciation of how landfill sites may affect water quality is given in paragraph 3.12 et seq. The same general considerations apply equally to the landfilling of difficult wastes. Any liquid waste deposition will feature in the overall water balance of a landfill, (see paragraph 3.33) contribute to leachate formation and hence, to the polluting potential of leachate. The polluting potential will generally be reduced by the attenuation processes described in paragraph 3.27 et seq.

7.16 Particular interest is attached to the potential impact of codisposal landfills since many of the components in an industrial waste will undergo some form of attenuation within a landfill when codisposal is practised. Freshly deposited household wastes have been found to possess very different attenuating properties to those exhibited by mature wastes. For many difficult wastes, codisposal with mature (that is 1 to 5 years old) household waste has been found to be preferable if optimum attenuation is sought. Experience and research over the past decade has demonstrated that, with proper management, attenuation processes can act such that the codisposal of difficult wastes does not result in unacceptable environmental pollution; the polluting potential of the leachate should be no greater than that from household or similar waste. However, it must also be recognised that the detailed behaviour of individual wastes, let alone that of complex mixtures within the physico-chemical and biological environment of a landfill is not as yet fully understood. Some attenuation processes such as those involving pH dependent reactions, ion exchange and adsorption are to some extent reversible. Any codisposal landfill operation must therefore ensure that conditions which could give rise to undesirable reactions are not allowed to develop, thus preventing the sudden release of pollutants. For example, the disposal of chelating agents or acidic wastes into landfills which have received significant amounts of metallic sludges.

7.17 The composition of leachate from several codisposal sites is shown in Table 3.23B and also shows typical figures for household wastes leachate composition. The various factors relating to the treatment of leachate, as outlined in paragraph 3.44 et seq, apply also to the treatment of leachate from landfills taking difficult wastes. Where leachate from a difficult waste landfill is discharged to sewer its possible effect on the sewer fabric and sewage treatment process may be the limiting factor on the volume and strength that can be accepted. Water authorities will take this aspect into account when imposing consent conditions.

General environmental effects of landfills accepting difficult waste

7.18 The general environmental effects of landfills taking difficult wastes — that is, the effects on landscape, ecology and the local community — should generally be no different from those of landfills taking only household and similar wastes. The principles and practices described in paragraphs 3.89-3.102 should therefore be followed.

7.19 Nevertheless, it must be accepted that landfills receiving industrial waste will be perceived as constituting a greater risk to health and safety in the locality concerned than those that do not. It is correspondingly even more important therefore, that operators of difficult waste landfills should take all possible steps to establish and maintain good relations with the local communities.

Treatment of difficult wastes prior to landfill disposal

Introduction

7.20 Some form of pretreatment will be necessary to render certain types of difficult waste suitable for landfill disposal. This may be because of its inherent reactivity, incompatibility with other wastes or handling difficulties due, for example, to a high liquid content. Pretreatment of waste is often carried out prior to its delivery to a landfill site. Occasionally, where suitable facilities are available, pretreatment may be undertaken at the disposal site.

7.21 Pretreatment of waste should be considered prior to landfilling to achieve:

(a) bulk reduction, eg dewatering of sludges;

(b) reduction of hazard potential when handling or transporting to a landfill; or

(c) conversion of the waste to an acceptable form for landfill disposal.

7.22 Pretreatment usually involves the use of physical or chemical processes such as phase separation, bulk reduction and solvent or solute separation. Chemical treatment may include neutralisation, oxidation and reduction, precipitation, or may aim to change the physical form of the waste, for example, through solidification or encapsulation.

7.23 Pretreatment of waste, particularly where special-

ised chemical treatment is required, will inevitably result in increased costs to the waste producer. Every effort should be made therefore to minimise waste disposal costs by reclamation, recycling or by volume reduction at the point of production to reduce transport costs.

Waste treatment by physical means

Bulk reduction of solid wastes

7.24 The bulk reduction methods applied to household waste, such as compaction and pulverisation are, in principle, suitable for many difficult wastes. However, their use can, in certain circumstances, lead to considerable difficulties; for example the generation of hazardous dusts or the release of hitherto contained corrosive materials. Incineration or drying followed by landfill of the residues is another way in which bulk reduction of difficult solid wastes may be achieved albeit at high cost.

Bulk reduction of sludges

7.25 Controlling the quantity of liquid waste discharged to landfill is important (see paragraph 7.15). Dewatering may render sludges more suitable for codisposal with domestic waste and should be considered.

7.26 Normal methods used for solid-liquid separation such as settlement, filtration, centrifuging and drying may all have a role to play in the pretreatment of sludges prior to landfill disposal. It may be possible to discharge the separated aqueous phase to a sewerage system but this will require the consent of the local water authority and a charge will be made. It should be noted that the removal of an aqueous phase will produce a solid containing higher concentrations of chemical constituents and could change the status of the sludge to make it a 'Special Wastes' as defined by regulations made under Section 17 of the Control of Pollution Act 1974 (see paragraph 2.68 et seq). Disposal licence conditions, may limit the quantity of an individual waste species which can be accepted at a site, having regard to the quantity of other wastes received. Consequently, the number of landfills which can accept the waste may be restricted.

7.27 Settlement of wastes using tanks provided with overflow facilities is the most simple dewatering option. Typically this can produce a settled sludge with a solids concentration of around 3 per cent weight/volume (w/v), which may be further increased by the addition of coagulation and flocculating agents. Lagooning, which is widely practised for large quantities of bulk wastes such as slurried Pulverised Fuel Ash, allows both settlement and evaporation to take place. Loss of volatile components, such as ammonia, may also occur from lagoons containing chemical waste slurries. Bottom solids concentrations are typically in the range 10 to 15 per cent w/v. If the lagoons are periodically allowed to dry out, a much higher solids content will result. In such instances physical stability aspects must be considered.

7.28 Filtration or centrifuging may be used to improve the solids content of a sludge or to separate more intractable mixtures. The waste slurry may be filtered through sand and gravel beds or by using various types of proprietary filters. Centrifuging can achieve relatively rapid separation and operate with wastes at elevated temperatures. Three phase systems, for example, oily sludges, may be centrifuged to produce three separate phases; a layer of oil for recovery, a layer of relatively clean water and a residual sludge.

7.29 Bulk reduction of sludges and slurries can also be achieved by drying on sludge drying beds or in specially designed driers. The former offers the cheaper option. Even when afforded protection from direct rainfall, performance is likely to be variable in the absence of additional heat. A final solids content of about 30 per cent weight/weight (w/w) is generally regarded as a practical limit for sludge drying beds. Unless low cost heat is available, forced drying is unlikely to be economical.

Bulk reduction of liquids

7.30 The evaporation and drying processes used to reduce the bulk of sludges can also be applied to liquids. In addition, freezing or chilling has been used. Other potentially useful techniques for separating solutes from solvents and some two phase systems include various membrane processes such as dialysis, electrodialysis, reverse osmosis and ultrafiltration.

7.31 Methods available for separating miscible liquids include distillation and solvent extraction. Distillation is widely used in industry for solvent recovery, but is probably too expensive for complex waste mixtures unless an adequate quantity of high value material can be recovered. Currently, solvent extraction has only limited applications for wastes, particularly those destined for landfill.

7.32 Treatment in settlement tanks with overflow of the supernatant may also be used for the separation of immiscible liquids, such as oil-water mixtures. The process can be aided by the addition of emulsion breakers.

Waste treatment using chemical processes

7.33 Many industrial wastes are given some form of treatment, at source, by the waste producer. Likewise many chemical reactions could be utilised to treat wastes but for various reasons, in practice comparatively few have been adopted. In many instances wastes are ill-defined heterogeneous mixtures which makes the design and operation of a processing unit difficult. The high cost of suitable reagents often makes their use economically unattractive. Restrictions on the presence of excess reagent, together with reaction products, in treated waste streams may be unacceptable in sewers. Chemical treatment processes can increase the final volume of wastes requiring disposal. Some of the more commonly used chemical treatment processes are considered below.

Neutralisation

7.34 Large quantities of high strength acidic or alkaline wastes are not acceptable for landfill since their corrosive properties or reaction potential exceeds the buffering capacity of the waste, within which they are to be deposited, to accommodate them. The dilution or neutralisation of large volumes of acidic and alkaline wastes is therefore normally necessary. While these wastes may be mixed in proportions to give a neutral pH product, extreme care must be exercised in doing so. Moreover, it is unlikely that exactly equivalent quantities of acidic and alkaline wastes will be available at the treatment plant at any one time. Additional neutralising agents may therefore be required. The addition of neutralising agents such as the addition of limestone to acidic wastes, will generate large volumes of carbon dioxide.

7.35 Lime and slaked lime are the alkalis most commonly used to neutralise acidic wastes but have the disadvantage of giving rise to large quantities of sludge for subsequent disposal. Sodium hydroxide, although more expensive has the advantage of minimising sludge formation and of being easier to control than lime or slaked lime. Often the neutralised products are soluble and therefore may be more easily handled. Sulphuric acid is widely used to neutralise alkaline wastes but can give rise to intractable sulphate slurries or solutions. Hydrochloric acid, although more expensive, has the advantage that little sludge is produced.

7.36 Neutralisation is normally carried out in a special plant equipped with settlement tanks, clarifiers, sludge filters and reagent feed mechanisms. Simple lagoons such as clay basins have also been used in which the neutralised sludge is allowed to settle and the supernatant liquor pumped away for disposal.

Precipitation

7.37 Precipitation techniques are mainly used to treat liquid effluents by reducing the concentration of the dissolved materials by coprecipitation or adsorption processes. The aim of precipitation is to produce a sludge which can be landfilled and a supernatant of sufficient quality for direct discharge to sewer or surface waters. However, the efficiency of the precipitation process is significantly reduced when the metal ions in solution are chemically bound as complexes. For example, hexavalent chromium compounds will not precipitate as a direct consequence of neutralisation. Many sludges contain relatively high concentrations of metals. Their presence should be taken into account when codisposed with household waste (see paragraph 7.62). Acidic conditions could result in dissolution of precipitated sulphides already present in a landfill.

Chemical oxidation and reduction

7.38 Chemical oxidation processes are sometimes used in the treatment of wastes prior to landfill, but reagent costs can be high. The reduction process most widely practiced in waste treatment is the reduction of hexavalent chromium compounds to the more environmentally acceptable trivalent state.

Sorption

7.39 Sorption is used to treat liquid waste streams by removing waste constituents from solution onto a solid support. Suitable reagents for sorption generally have large surface areas and are therefore inefficient for treating wastes containing significant quantities of particulates.

Solidification

7.40 Solidification is a process whereby a liquid or sludge is mixed with various materials to produce a solid product in which the waste material is more or less evenly distributed throughout the mass. The solid reduces the rate at which chemical species present in the waste can be leached out. Four main types of solification process have been evaluated and proposed to treat wastes. These are:

(a) cement based processes;

(b) lime based processes;

(c) thermoplastics and

(d) organic polymers.

Except for some very special purposes, only the first two processes have been commercially developed and exploited.

7.41 The mechamisms by which waste is retained in the solidified product matrix are unclear and may possibly include precipitation of insoluble hydroxy-species, the formation of complex calcium silicates, adsorption onto the solid matrix and physical enclosure of material within a crystal lattice. The reduction in leaching rates achieved may result from physical and chemical retention in the solid matrix and the high pH of the products. The proportions of reagents added to a particular waste may be arranged to provide a product with predictable leaching characteristics.

7.42 Solidification processes are particularly suited to the treatment of inorganic wastes and may be the preferred route when significant quantities of toxic metals such as mercury, cadmium, arsenic, lead and antimony are present in the waste. The presence of particular organic materials may impair setting of the product so that their application in the treatment of organic wastes is restricted.

7.43 Deposition of solidification process wastes in the UK is usually a monodisposal operation. Properties required of a solidification product are that it; sets fairly rapidly, has a low permeability, a slow rate of leaching and remains stable in the long term. Indications to date suggest that the properly formulated product is stable and that land restoration will not be adversely affected by settlement or product decomposition. Although solidification process products possessing high compressive strength and low permeability can be produced, their application as building materials, or to encapsulate untreated wastes, have not as yet found widespread use.

Loading rates and compatibility

Chemical interactions between wastes

7.44 When two or more wastes are to be deposited at the same location on a landfill site it is the responsibility of the site operator to ensure that the concentrations of reactive species in the wastes are sufficiently low that they may be safely deposited together. To achieve this it is essential that the composition of all wastes delivered to a landfill site are known before they are accepted at the site. Steps which must be taken to prevent the co-deposition of incompatible wastes include prior knowledge of the constituents in a waste; the direction of non-compatible wastes to separate parts of the site; the clear labelling of trenches for liquid waste reception and competent supervision of the discharge of wastes.

7.45 Undesirable reactions which can occur when mixing incompatible wastes include:

(a) the generation of heat by chemical reaction (exothermic reaction) which in extreme cases may result in fires or even explosions, eg alkali metals, metal powders;

(b) the generation of toxic gases, eg arsine, hydrogen cyanide, hydrogen sulphide;

(c) the generation of flammable gases, eg hydrogen, acetylene;

(d) the generation of gases such as nitrogen oxides, carbon dioxide, sulphur dioxide, chlorine, and

(e) dissolution of toxic compounds including heavy metals, eg complexing agents, chelates.

In the UK only limited assessment work on the compatibility of general classes of industrial waste has been undertaken. Consequently most of the available information is based on United States data. A simplified version of an extensive study in which the reactions between the 12 most common categories of hazardous wastes are compared with one another, is presented in Figure 7.45.

Figure 7.45 COMPATIBILITY OF SELECTED HAZARDOUS WASTES

Legend

E	Explosive
F	Fire
GF	Flammable Gas
GT	Toxic Gas
H	Heat Generation
S	Solubilisation of Toxins

#	Waste Category	1	2	3	4	5	6	7	8	9	10	11
1	Oxidising Mineral Acids											
2	Caustics	H										
3	Aromatic Hydrocarbons	H F										
4	Halogenated Organics	H F GT	H GF									
5	Metals	GF H F	S									
6	Toxic Metals	S										
7	Sat Aliphatic Hydrocarbons	H F										
8	Phenols and Cresols	H F										
9	Strong Oxidising Agents	H		H F		H F		H				
10	Strong Reducing Agents	H F GT			H GT		S		GF H	H F E		
11	Water and Mixtures containing Water	H			H E						GF GT	
12	Water Reactive Substances											

Row 12 (Water Reactive Substances): Extremely reactive, do not mix with any chemical or waste material

7.46 Very little information is available on adverse reactions involving three or more wastes although codisposal involving several waste streams has been successfully undertaken on the basis of accumulated experience. It is important that where such experience is lacking and where adverse reactions are possible, small scale testing should be undertaken by competent staff to establish the compatibility of wastes before they are landfilled.

Waste-liner compatibility

7.47 The lining of landfills has been discussed in Chapter 4 where it is stressed that the integrity of the lining material is vital. However, it has been recognised recently that certain wastes may adversely affect the performance of a liner. It is important therefore to ensure compatibility between the liner material used and the wastes to be deposited. In view of the uncertainties that exist, especially on the long-term behaviour of liners, certain precautions appear desirable in controlling the deposition of difficult wastes, as described below (see also paragraph 4.57 and Appendices 4a and 4b).

Clay and bentonite liners

7.48 In the light of very limited current knowledge it is suggested that incompatible difficult wastes should not be deposited adjacent to in-situ clay which is less than 10 metres thick. Liner materials based on bentonite exhibit properties which make them compatible with a wide range of chemicals. Unlike many synthetic liners, bentonite is not adversely affected by oils and organic solvents except at high concentrations ($>$50%) although of course these would not be landfilled. Large volumes of acids and alkalis should not be deposited at bentonite lined sites. There is no known experience of a codisposal site having been lined with bentonite.

Polymeric membranes

7.49 Various polymeric membranes are available which, though compatible with inorganic wastes may, to varying degrees, be susceptible to attack by organic chemicals. Such membranes should not be used therefore to line sites that are expected to receive organic chemical wastes until their compatibility with the wastes has been established. Difficult wastes should not be deposited within 5 metres of an artificial liner.

Loading rates

7.50 There is a need to limit and control the quantities of wastes deposited at a landfill site in order to maintain appropriate proportions between various types of waste. Codisposal sites in particular depend on careful regulation of waste types and quantities deposited to ensure that attenuating processes do not become overwhelmed.

7.51 For difficult wastes to be codisposed with household or commercial wastes their rate of deposition will be limited by the attenuation capacity of the latter. Although general guidelines can be given, the optimum loading inevitably will be site specific. Operating experience and effective monitoring of the conditions within a landfill provide the best guidance for setting appropriate loading rates.

7.52 Loading rates are normally quoted as a quantity of waste which may be deposited either in a given time period (eg as tonnes per week), or with a given quantity of another waste (eg as tonnes per tonne of household waste). In the latter case the quantity of waste already deposited is of prime interest, rather than the daily quantity of household waste arriving at the gate. It is essential therefore, that there is a sufficient volume of household waste already deposited to assimilate the proposed addition of industrial wastes.

7.53 Most codisposal landfills are licensed to take a wide range of wastes containing a large number of potentially polluting components and it is necessary therefore to identify those constituents in wastes which will requires individual monitoring. It is considered unnecessary to set loading rates for every constituent since, not only would the analytical requirements be onerous but could not be technically justified. Similarly it is impracticable to measure precise inputs of particular components in all waste consignments received. Loading rates should take such imprecision into account. Nevertheless, at all sites there will be certain chemical species in wastes received that must be carefully estimated, regulated and monitored.

Guidance on the loading rates for specific wastes

7.54 Experience to date and research have not yet been able to establish fully the potential for codisposal. The ongoing practice of landfilling industrial wastes will inevitably increase the amount of monitoring information available to assist in setting loading rates appropriate for particular types of site. The following paragraphs provide information additional to that contained in the current edition of papers in the Waste Management Series (see Appendix 7a).

Liquid wastes

7.55 In general, it is the volume of liquid wastes, rather than the capacity of the site to attenuate constituents in the waste which controls the addition of liquids to landfills. For management purposes water balance calculations must be used throughout the lifetime of any landfill site accepting liquid wastes. The quantity of liquid waste deposited will be sensitive to, and may be temporarily controlled by, other events which contribute to water balance, such as above average rainfall, or other wastes having sufficient absorptive capacity not being available. The effects will be most noticeable at containment landfills. At such sites it is therefore desirable to control the addition of liquid wastes according to the results obtained from monitoring the observed liquid levels in the site (see also paragraph 7.100).

Acid wastes

7.56 Waste Management aspects of acid and alkali wastes are at present being considered by the Department with a view to publication of a Waste Management Paper in 1986. The following guidance should therefore be regarded as provisional until publication of the Waste Management Paper. Concentrated acids should not be directly landfilled since they are extremely corrosive and if landfilled have the potential to cause fires and produce toxic gases by chemical reaction. When for technical reasons landfill is presently the best practicable option the acid should, as a minimum, be diluted prior to deposit. Nevertheless, household and similar wastes have a significant capacity to neutralise acids; this capacity is normally maintained until degradation of the waste is almost completed.

7.57 Codisposal of acidic wastes should only be at a rate which will maintain a near neutral pH value in any leachate produced. In addition four other effects must be considered. First, acid waste can dissolve metals present in other waste which will require subsequently to be attenuated to acceptable levels. Second, the quantity of liquid represented by any acids deposited must be within the sorptive capacity of the landfill. Third, acid inevitably will have a toxic effect locally on the microbiological population, although evidence suggests that recovery is relatively rapid. Fourth, acids can react with other materials in a landfill to produce toxic gases. Such effects should be monitored in areas adjacent to those receiving acids and input quantities regulated accordingly.

7.58 The codisposal of acidic wastes into trenches excavated into previously deposited depends upon a sufficient quantity of absorptive waste being available for neutralisation as the acid waste moves away from its immediate point of disposal. Mature household waste, (1 to 5 years old) usually affords good buffering capacity. When trenches are used their lifetime will vary. Subsequently, where vertical flow can occur, at sites located on permeable strata, a lifetime of several months can be expected, while at sites having low permeability base material and hence only lateral leachate migration can occur, there is evidence that the neutralising capacity may not be exceeded for several years.

7.59 For direct disposal to landfill the concentration of hydrochloric and sulphuric acids should not exceed 20%, nitric acids 5% and chromic acids 5% (as chromium VI). Chromic acids must be neutralised or diluted to pH 4 to attenuate its uncontrolled reaction with organic matter. Mixed acids require special consideration since their reactivity is generally greater than that of the individual acid alone.

7.60 The loading rate for sulphuric acid should be limited to a maximum of 20 kg of acid (by weight of pure acid) per tonne of mature household waste. When calculating initial loading rates at sites located on permeable strata it will be necessary to ensure that there is sufficient waste directly below and immediately adjacent to the disposal trench. Where lateral spread of waste acid is expected, larger amounts of household waste will be available for neutralisation. Under these conditions acceptable quantities may be much larger but this can only be established by careful monitoring within the site in relation to the permeability of wastes previously deposited, the available absorptive capacity remaining around the trench and the blinding effect of precipitation reactions occurring at the base of the trench.

7.61 In the absence of liquid movement laterally, a loading rate of no more than 5 kg of hydrochloric acid per tonne of mature household waste is suggested since hydrochloric acid is particularly effective at mobilising metals. The mobilisation of heavy metals is a good indicator of the effect of acid codisposal when monitoring sites. Zinc and nickel are typically the first to appear in leachate. The degree of mobilisation is likely to depend as much on the competition by other ions in the waste acid as on the effect of the acid itself. Dilute solutions of hydrochloric or sulphuric acid ($\leqslant$1%) are less aggressive in a landfill and loading rates will probably be controlled by the quantity of liquid rather than any chemical reaction.

Heavy metal wastes

7.62 In terms of environmental impact potential or quantity the principal heavy metals of concern are cadmium, chromium, copper, lead, nickel and zinc. They are often found together in a single waste stream. Wastes which may contain heavy metals include hydroxide sludges, ashes, slags and some paint pigment wastes. Such wastes are almost always landfilled in a water insoluble form and care is needed to ensure that they remain so in the landfill environment (see paragraph 7.44 et seq).
Excluding cadmium, heavy metals are frequently considered together when setting discharge consents to sewer for leachate. The concentration of zinc in leachate from fresh refuse (generally about 10 mg/litre) may require it to be considered separately. Cadmium should also be considered separately in view of EC groundwater directive (see Appendix 7b). The loading rates for heavy metal wastes will depend on their leachability potential and on subsequent sorption/desorption and precipitation reactions.

7.63 The concentration of heavy metals found in household and similar wastes can vary widely. Codisposal of heavy metal wastes should not be allowed if it is likely to increase their concentrations beyond those observed in leachate derived from fresh refuse; it is best carried out using household waste more than 6 months old (see Table 3.23 A and B).

7.64 Limited evidence suggests that an initial loading rate of up to 100g of soluble chromium, copper, lead or zinc per tonne of mature household waste is unlikely to produce a significant change in leachate concentrations 3 metres distance from the heavy metal waste. The likelihood of insoluble metal compounds becoming soluble also needs consideration. In the absence of field data demonstrating the contrary it should be assumed that nickel will not undergo significant attenuation other than by dilution and dispersion. The codisposal of heavy metal wastes may inhibit microbial organic degradation processes within the waste mass. Resistant populations can, given time, be re-established. It is desirable therefore, to monitor the microbiological activity in such instances.

Arsenic, selenium and antimony

7.65 Provided it can be demonstrated that the concentration of arsenic does not exceed 10 mg/litre in leachate from a codisposal site, industrial wastes containing up to 1% total arsenic may be acceptable for disposal. In wastes containing soluble arsenic compounds pretreatment will usually be required to ensure that the concentration of arsenic in leachate from the site does not exceed 10 mg/litre.

7.66 Isolated small quantitites of concentrated arsenical waste arisings containing less than 0.5 kg of total arsenic may usually be codisposed without pretreatment. However, such wastes should be covered immediately by at least 2m of refuse, and, except for arsenic sulphides, should where practicable be mixed with lime. The production of nascent hydrogen which can lead to the generation of arsine gas must be obviated in all landfill operations involving arsenical wastes. Additional information on arsenic bearing wastes is available in Waste Management Paper No 20.

7.67 Selenium and antimony bearing wastes should be treated in a similar manner to arsenic since their chemical behaviour is generally analogous. However, they do not form stable hydrides and there is therefore less likelihood that they will present a problem. Nevertheless, loading rates for selenium and antimony should be based on similar criteria to those used for arsenic.

Mercury

7.68 Before establishing loading rates for wastes containing mercury in landfill, it is essential to know the chemical form in which the mercury is present. Although all mercury compounds have the potential to accumulate in tissue, organic and particularly alkyl mercury compounds are the most toxic. Loading rates should therefore be set to prevent elevated concentrations of mercury occurring in leachate and consideration given to the possibility of mercury vapour being released.

7.69 Where significant quantitites of mercury-bearing wastes are involved, their disposal to landfill should be restricted to sites where little or no horizontal leachate flow occurs. A special disposal area should be designated for wastes containing more than 20 mg/kg inorganic or elemental mercury or more than 2 mg/kg organic mercury. Such an area will require special restoration treatment and its afteruse may be restricted. For non-specific mercury bearing wastes the loading rate should not exceed twice the average concentration of 2 g/tonne found in household and similar waste. Individual deposits of solid waste containing more than 1 kg of mercury are not acceptable, unless distributed through large areas of uncontaminated refuse.

7.70 Such initial loading rates may not take full advantage of the efficiency with which soluble mercury is precipitated as mercuric sulphide in the anaerobic zone of a landfill. However mercury bearing wastes containing viable concentrations of non-organic mercury should be recovered, and those containing more than 100 mg/kg organic mercury should not be accepted for direct landfill. Additional information is presented in Waste Management Paper No 12.

Phenolic wastes

7.71 Phenolic compounds such as phenol, cresol and xylenol are moderately soluble in water. Their presence at extremely low concentrations in waters used for potable water production will, on chlorination, impart unpleasant taste and odour to the water. The codisposal of phenolic wastes therefore requires particular care to prevent contamination of water used as a source potable water production. Phenolic wastes in landfill sites are attenuated by both aerobic and anaerobic degradation following sorption. Long residence time in the landfill and good contact between the waste and leachate is necessary to achieve significant attenuation.

7.72 Phenols have a greater affinity for organic materials than for water. To maintain the partition in favour of contact between phenolic and the organic fractions of codisposal waste and also delay leaching, the use of low permeability intermediate cover material should be employed. Landfill sites on low permeability strata and the use of discrete cells offer distinct advantages in this respect.

7.73 Leachates from fresh refuse have been found to contain total phenol concentrations of 1 to 10 mg/litre whereas leachates from older refuse contain about 1 mg/litre. Codisposal using refuse more than 6 months old therefore offers distinct advantages. To maintain concentrations within the 1 to 10 mg/litre range water infiltration should be minimised. Deposition of solid wastes containing phenols may be covered with a minimum of 300 mm of clay or similar low permeability material to minimise water infiltration into the waste. Phenolic wastes should not be deposited where perched water tables are likely to be a problem. Liquid phenolic wastes are more likely to be acceptable at containment landfill sites but in any case only in very small quantities or at low concentrations.

7.74 The loading rate of phenolic wastes to landfills should not normally exceed 2 kg of total phenols per tonne of refuse, unless site specific data confirm that biodegradation of phenols is particularly effective. Acceptable attenuation cannot be inferred by an observed low initial rate of leaching. There is evidence that concentrations of 1 mg/litre phenol are reduced quite rapidly in the unsaturated zone outside a landfill site in strata possessing slow intergranular flow. Attenuation is thought to be by anaerobic degradation which may occur within many landfills containing only household waste but it is not sufficient to place reliance on this mechanism always occurring.

Oily wastes

7.75 Oily wastes typically occur as three different types: free oils, industrial oil-water emulsions and oily mousses which is usually a solid material such as that derived from marine oil spills. Attenuation of oils in landfills is largely through sorption by solid waste. Degradation rates of the sorbed material are extremely slow. Aerobic conditions using land farming techniques have been shown to be more efficient in degrading oils. For these reasons wastes containing oils should go to landfill only when other means of disposal are not practicable but it is recognised that landfill may be necessary for oil removed from beaches after an oil spill at sea. Recovery of oil from waste, for reuse or use as a fuel, should be undertaken whenever possible, and its incineration as a source of heat when this is not so. Oils which cannot be recovered or incinerated are likely to be so because they have become contaminated with unacceptable materials. Loading rates must take into account the exact nature of such contaminants in addition to the oils.

7.76 The concentration of mineral oils found in leachate from household waste is typically 10 mg/litre. Where oils are deposited at loading rates of not more than 2.5 kg of oil per tonne of well compacted refuse, the concentration of oil in the leachate is unlikely to exceed 10 mg/litre, after passage through a few metres of household waste.

7.77 Oil-water emulsions, such as cutting oils which also normally contain bactericides and additives to stabilise the emulsion, have been found to infiltrate landfills and to be absorbed by household waste in a similar manner to aqueous wastes. Subsequent breakdown of the emulsion allows absorption of the free oil by solid waste in the same way as with mineral oils, the aqueous phase becoming leachate. In order to keep oil concentrations in leachate within the range of those from household waste, loading rates should not exceed 40 kg of oil emulsion or 0.4 kg of oil per tonne of refuse. If disposal trenches are used, it should be noted that some emulsions, on breakdown, tend to blind the base and sides of trenches.

7.78 There is some evidence to suggest that containment sites are more permeable to oils than water. Site loadings must therefore be limited to such amounts that leachate concentrations typical of those from refuse alone are not exceeded. In addition, due to their stability, oils are unsuitable for deposition within the top layer of refuse at codisposal landfill sites. Further information on the disposal of mineral oil wastes is given in Waste Management Paper No 7.

Pesticide wastes

7.79 Concentrated pesticide wastes are unsuitable for landfill disposal and should be destroyed by high temperature incineration. Wastes containing low concentrations of pesticides, eg tank washings, may be acceptable for codisposal although such practices should be closely monitored.

7.80 Waste Management Paper No 21 recommends maximum concentrations of 10 g/m^3 or 20 mg/kg of pesticide active ingredient in the waste after deposition. Leaching of pesticides appears to be largely controlled by their water solubility so it is possible that these concentrations could be increased for some pesticides having low solubility. However, evidence will be required by the licensing authority to show that such materials can be deposited in an environmentally safe manner. Deposition should take place in trenches excavated in mature household waste. Immediate backfilling is recommended for solid waste, although trenches used for liquid wastes may be left open until an acceptable loading has been achieved.

Polychlorinated biphenyl (PCB) wastes

7.81 PCBs are persistent in the environment and there is no evidence that they degrade significantly even under anaerobic conditions within a landfill. They are relatively immobile and their water solubility is low, typically 0.01 to 0.001 mg/litre. Sorption of PCBs is likely to occur on solids with high organic content. The disposal of anything but minimal amounts of PCBs dispersed through a landfill is unacceptable. Recent guidance from the Department has indicated that 20 mg/kg PCB in waste after deposition, in admixture with bioactive waste within the body of a landfill, should be regarded as a maximum concentration. Additional information is available in Waste Management Paper No 6.

Solvent wastes

7.82 Solvent wastes which burn unsupported at 40°C or below should not be landfilled. Other solvents of lower flammability and particularly those mixed with other wastes can be landfilled only in small quantitites. Guidance on the disposal of solvent wastes is given in Waste Management Paper Nos 9, 14 and 15.

Acid tars

7.83 Although Waste Management Paper No 13 indicates a theoretical suitable loading rate for acid tars, in intimate admixture with household waste, it is impracticable to meet the conditions that allow this loading rate to be achieved. Acid tars arise mainly from three industries, the production of metallurgical coke, the manufacture of highly purified mineral oils such as the white oils used by the pharmaceutical industry and the re-refining of used oils. They all contain a mixture of often highly viscous thixotropic organic material and sulphuric acid of variable but usually high concentration and in some cases even oleum. Acid tars do not degrade to any significant extent in a landfill environment and when buried have a tendency to migrate upwards, leading to uncontrolled releases at the surface.

7.84 The landfilling of acid tars raises very difficult problems and it is doubtful whether such wastes should be landfilled at all; the matter is currently under investigation by the Department. Very few landfills are capable of safely accepting acid tars and none should do so without pre-treatment; as a minimum the free acid should be neutralised and any free oil separated before the residue is landfilled by codisposal.

Cyanide wastes

7.85 Cyanide can be present in waste in various chemical forms such as: 'simple cyanide', 'complex cyanide', 'cyanate', 'thiocyanate' and 'nitriles'. In all instances it is the ability of the waste to produce free cyanide which dictates the loading rate, in a landfill. The degradation mechanisms for cyanide are discussed in the second edition of Waste Management Paper No 8. In general terms disposal to landfill is recommended for only small quantities.

7.86 Cyanide wastes should be deposited in segregated areas of a landfill site where the pH will not fall below 5, thereby minimising production of hydrogen cyanide. Complexed cyanide wastes, such as soils contaminated with gas works wastes containing 'spent oxide' should be deposited at loading rates which take account of the concentration of free cyanide and other contaminants which are likely to be present such as phenols. Spent oxide typically contains 10 to 50 mg/kg free cyanide.

7.87 In general, cyanide waste should be deposited at a loading rate of no more than 1g of free cyanide per tonne of refuse. Further information is available in Waste Management Paper No 8. An exception may be wastes from the redevelopment of old gas works sites where significant quantities may require disposal during a limited period of time. In such cases it may be acceptable to exceed the recommended maximum loading rate provided such a deposit does not overwhelm the attenuation processes operating within a landfill which could lead to long term environmental problems. Containment landfills may offer considerable advantages for the disposal of these wastes.

Tannery sludges

7.88 The loading rate for the codisposal of tannery sludges depends both on the water content of the waste and chromium content. For a typical sludge solids content of 2% the loading rate should not exceed 1 tonne of sludge per 15 tonnes of refuse. However, dewatering of the sludge prior to landfill is preferable, the loading rate of the sludge being adjusted accordingly. Tannery sludges typically contain up to 3.5% chromium (III) in the dry solids. In practice the leaching rates of chromium from these wastes are very slow and elevated chromium concentrations are rarely observed, nevertheless, monitoring for the presence of chromium in leachate is advisable. Waste Management Paper No 17 gives detailed information.

Monitoring

7.89 Monitoring is essential at all sites where codisposal is practiced. The objective of monitoring is to build up a picture of the general behaviour of a site. The distribution of key species in leachate within a site, together with physical properties such as temperatures and water levels, indicate the general state of a landfill and show the effect of wastes deposited. Initially, wastes deposited should be limited to household and other bioactive general waste to allow the monitoring network discussed in Chapter 3 to become established and to ensure that a sufficient depth of sorptive waste is available before codisposal commences. The impact of the initial deposits of codisposed industrial waste can thus be monitored and provide information on appropriate loading rates for the site.

7.90 It is unlikely that the loading rates initially set will remain relevant for the lifetime of a site. The monitoring data, which should be interpreted and discussed between the site operator, water authority and disposal authority, will enable a clear picture of site behaviour and its attenuating properties to be established which, in turn, will indicate whether or not loading rates require adjustment.

Site operations

Introduction

7.91 The operation of a landfill site has been discussed in detail in Chapter 5. This section therefore deals with those additional procedures that may be necessary at sites taking difficult wastes.

Monodisposal

7.92 Monodisposal lagoons are best suited to large

arisings of dilute liquids or sludges. Monodisposal sites, especially lagoon sites, normally occupy several hectares. Consequently, large areas of land may be committed to this use for some considerable time. With proper management it is possible to minimise its impact on the environment to an acceptable level. In many instances such sites can be operated on the principle of progressive restoration, using a cell method which allows completed areas of the landfill to be made available for other uses as each cell is restored. Even during operations, sympathetic landscaping can reduce the level of intrusion and help reduce the environmental impact.

7.93 Monodisposal normally makes little or no use of attenuation processes within the landfill. The restricted concentration and chemical composition of the waste and its relatively uniform nature usually limits the physical, chemical and biological attenuation mechanisms that can operate. Care must therefore be taken to ensure that the disposal operation does not lead to long term sterilisation of land. This principle applies generally to all landfill operations.

7.94 The function of a monodisposal lagoon is to permit settlement of suspended solids from the waste; typically 99% of the solids can be expected to settle out in the lagoon. Lagoons are usually containment facilities, but slow seepage may occur at some sites. In most cases the aqueous supernatant is discharged from the lagoon to a watercourse under consent conditions agreed with the water authority. A small proportion of the effluent may be lost by evaporation. At some sites advantage is taken of the extensive surface area of a lagoon to assist in cooling hot effluent to near-ambient temperatures and evaporate minor constituents such as dissolved gases or vapours, eg ammonia, or low boiling point solvents it contains.

7.95 Solids which settle in the lagoon are not usually removed except in instances where space is limited or there is a recovery option. In such cases a lagoon is effectively operated as a simple treatment plant, in which slurries are thickened by settlement to reduce the transport costs for final disposal. Generally, wastes that are lagooned contain only low concentrations of solids so that the rate of infilling is slow; thus lagoons have long lives.

7.96 Monodisposal lagoons are typically constructed on flat ground. The retaining walls are usually formed by preparing a shallow excavation along the line of the wall the material excavated being used to form the wall. Alternatively natural materials (clay etc) or waste ash and clinker may be used. No attempt is usually made to render the base of lagoons impermeable. Indeed, many older lagoons incorporate underdrainage. Some downward flow of the aqueous liquor may thus be expected to occur and account should be taken of this in assessing the possible pollution of ground water. Care is also needed to maintain the integrity of the retaining walls to prevent their failure.

7.97 One vital aspect in lagoon management is to ensure that discharges and drainage are maintained within appropriate water authority consents. Control over lagoon operation may be affected by the location of the inlet and exit points and the depth of liquor in the lagoon. In shallow lagoons windy conditions can induce wave motion, resulting in resuspension of solids and a deterioration in outfall water quality. The effect can be minimised by increasing the depth of liquor on the bed. It should be noted that lagoons containing impounded water above surrounding ground level in excess of 25,000 m^3 will be subject to the provisions of the Reservoirs Act 1975.

7.98 While a landfill site can be provided with a domed profile on completion, when a lagoon becomes full an extensive area, which is either completely flat or more often depressed due to contraction as the wastes continue to dry, will be produced. The effects may be experienced even though the deposits have been covered. Drying of the bed is critically dependent upon the drainage characteristics of the lagooned waste and the permeability of the walls and base of the lagoon. Coarse deposits may dry out relatively rapidly but, fine deposits may take decades to reach a state in which they can be safely walked upon or take vehicular movements. While lagooning is capable of raising low-lying land to a more usable level, there may be problems with its afteruse.

Codisposal

7.99 The way in which codisposal is carried out depends on the nature of the waste and the requirement to achieve satisfactory loading rates. The following paragraphs give general guidance but it should be recognised that codisposal operations are site specific.

Codisposal of liquid wastes

7.100 Before any liquid wastes are accepted for codisposal, the capacity of the site to receive them should be assessed by making a water balance calculation (see paragraph 3.33 et seq). Specific areas of the site filled with medium density household or similar wastes (0.6-0.7 tonnes/m^3) should be designated for liquid disposal to facilitate its absorption. Highly compacted wastes may have a lower capacity for absorbing liquids. Deposition of liquids into baled wastes is inadvisable, since high density bales limit the quantity of liquid which can be absorbed as well as encouraging fissure flow. It is preferable to deposit liquids into wastes which are at least 6-12 months old.

Liquid reception facilities

7.101 Liquid wastes are usually taken directly to their point of disposal on a landfill or sometimes stored temporarily in tanks or sumps at the site entrance; from which they can later be pumped to trenches or lagoons on the landfill. Precautions to ensure that only compatible liquids are allowed to be mixed in temporary storage tanks is essential.

Methods for codisposal of liquid wastes

7.102 Four methods are available for the application of liquid wastes to landfill: trenching or lagooning, injection, spraying and surface irrigation.

7.103 *Trenching and lagooning* When excavating trenches or lagoons, special care must be taken to ensure that the area of the site has not previously been used for the deposition of wastes which will be incompatible with incoming wastes. Other possible hazards should also be considered; for example, no excavations should be allowed where asbestos has previously been deposited. All excavated material should be dealt with in such a way that no nuisance is created, for example, by windblown material or odours. Trenches and lagoons should always be clearly marked and fenced. Only trenches should be used for the codisposal of odorous wastes and the possibility of nuisance reduced by covering the trenches with such materials as railway sleepers, plastic, corrugated iron or purpose built covers.

7.104 The distance between trenches should be at least 5 metres and they should be at least 10 m away from the site boundary. At lined containment sites it is necessary to leave sufficient underlying waste and take other measures needed to protect the integrity of the liner. Precautions should be taken to ensure that the sides of trenches are stable and fencing provided to prevent unauthorised or inadvertent access. Edge protection may be necessary where access to trenches is required.

7.105 The aim of trenching is to provide sufficient internal surface area to allow seepage into the waste mass already deposited. Various designs have been used (see Figure 7.105). The number and disposition of trenches required will depend both on the quantity and type of incoming wastes. Separate trenches should be allocated to different types of waste. The liquid levels should not be allowed to reach the top of a trench at any time, and surface waters should be diverted away from them. Fine material can accumulate on the base of a trench and retard seepage. This may be alleviated by providing a number of interconnected trenches using an overflow 'weir' system. Liquid codisposal operations should always be monitored.

7.106 Lagooning on a landfill site is the disposal of pumpable fluids into a bunded area on the surface of a landfill, or more usually into shallow pits excavated in waste already deposited. Where codisposal lagooning is practiced, care must be taken to ensure that the sorptive capacity of the underlying solid waste is not exceeded and disposal must be carefully controlled and managed to avoid chemical incompatibility. Liquids are dissipated by seepage into the waste and during summer months, by evaporation. Completed lagoons may have poor load-bearing characteristics, and should remain fenced until the land becomes stabilised.

7.107 *Injection* Liquid wastes can be injected below the surface of a completed landfill using facilities which will have been installed during the lifetime of a site. The facilities provided, for example, by installing perforated pipes or by building up columns of tyres, can extend from several metres above the base of the landfill up to the finished level for waste deposition. The number of 'wells' provided and the outlets at different depths to utilise fully the capacity of all the solid waste deposited will depend on prospective waste arisings, and the size of the site. Specific 'wells' should be designated for specific wastes.

7.108 The main advantage of injection into completed sites is that a more accurate assessment of the water balance can be made before liquid wastes are accepted. The quantity of incoming liquid wastes can therefore be regulated to ensure that the absorptive capacity of the landfill is not exceeded. Injection can also offer certain advantages for the disposal of odoriferous wastes.

7.109 *Spraying and surface irrigation* The spraying of liquid wastes over the surface of deposited waste on land-fill sites makes use of evaporation to reduce the volume and possibly aereal oxidation to replace the organic loading of liquid wastes, and can also possibly improve leachate quality by allowing further reaction with the solid waste mass to occur. Care must be taken to avoid pollution of surface water and excessive ponding. Spray drift and potential health hazards to operators also need to be considered.

7.110 *Codisposal of sludges* Sludges and slurries can be deposited in trenches in a similar manner to liquid wastes (see paragraphs 7.103-105) but rapid blinding may occur. Alternatively trenches can be excavated in waste already deposited in advance of a working face, the sludge deposited and immediately covered. Sludges may also be deposited with incoming solid waste provided that due regard is taken of water balance. Sludges should be deposited at the base of the working face and covered immediately with solid waste. Care should be taken to avoid running vehicles over the sludge.

7.111 *Codisposal of solid wastes* Generally, solid difficult wastes should be dispersed along the base of the working face and covered immediately. Such wastes should not be deposited within 5 m of a site boundary or within the first lift of waste deposited on a site. If difficult solid wastes are to be deposited at sites where thin layer techniques are employed, special areas must be designated away from the working face.

7.112 Impermeable wastes are not suitable for depositing in layers over large areas since they are likely to give rise to perched water tables and channelling of the leachate. Attenuation processes within the landfill are consequently likely to be inhibited under such circumstances.

Figure 7.105 CODISPOSAL TRENCH DESIGN

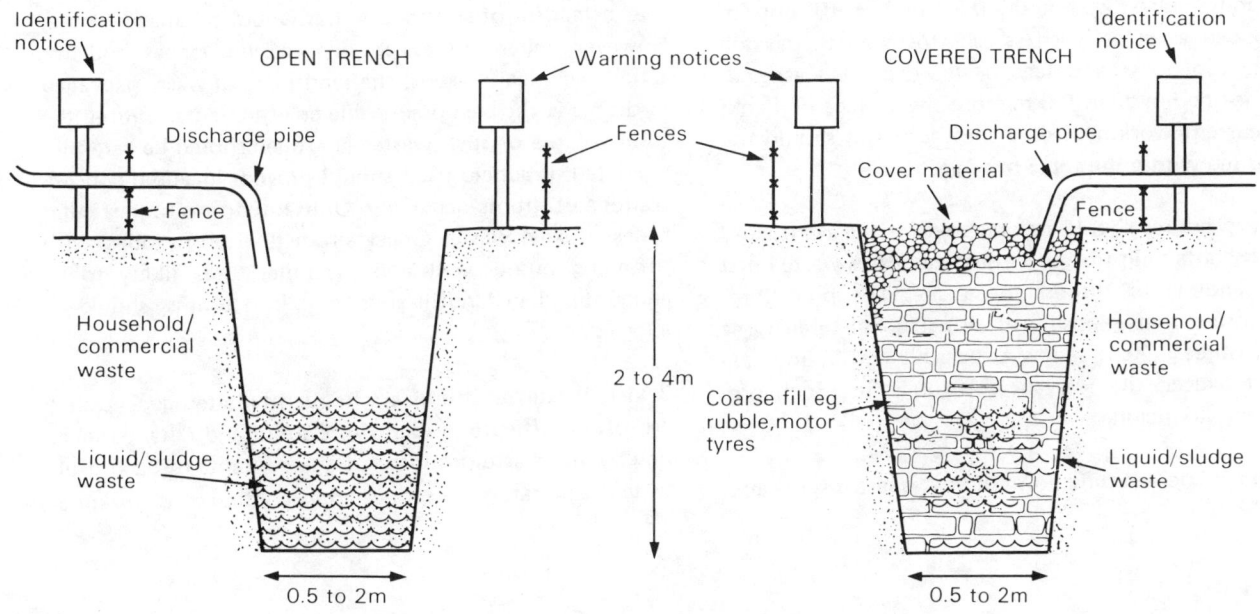

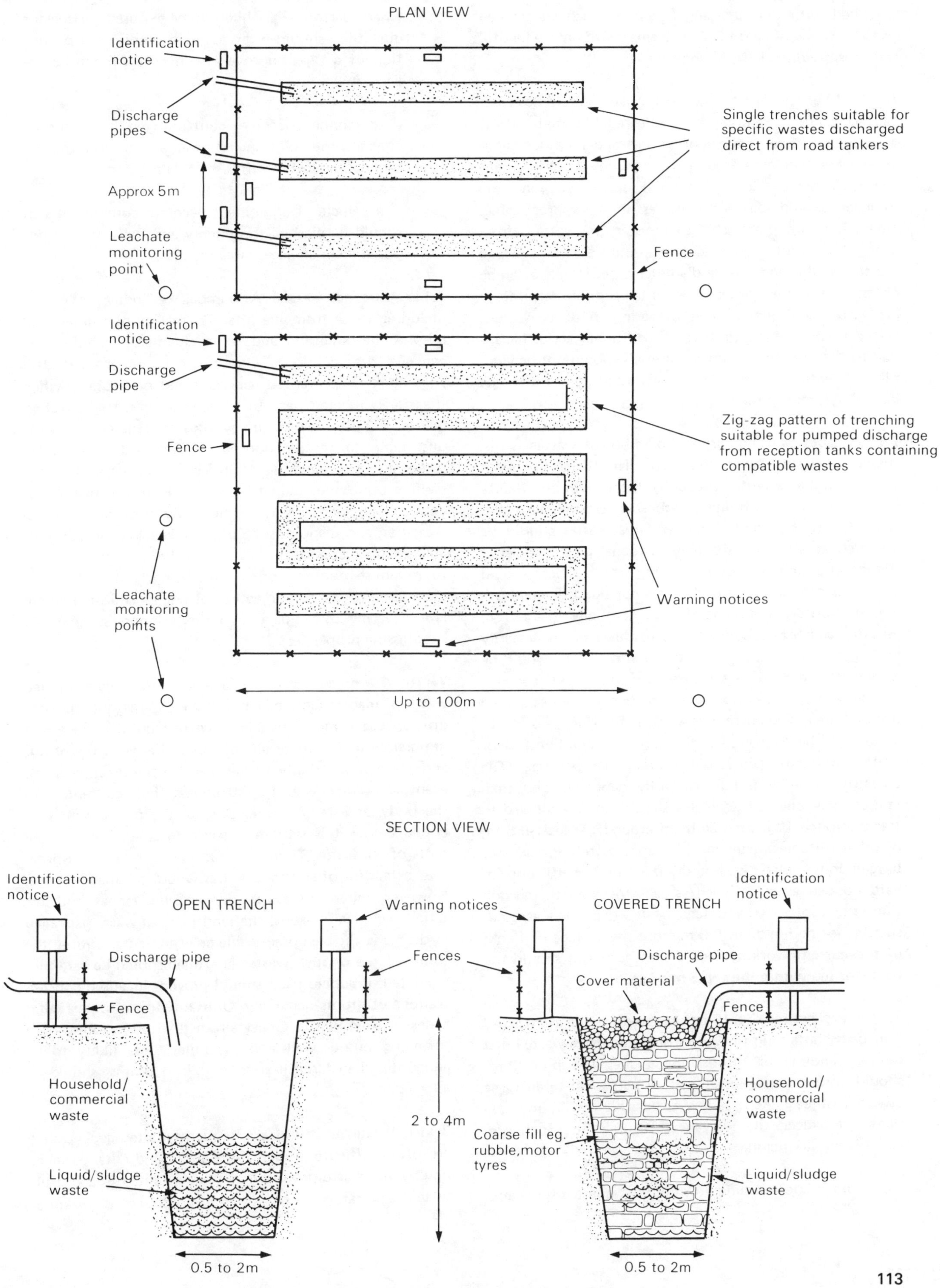

PLAN VIEW

Identification notice

Discharge pipes

Approx 5m

Leachate monitoring point

Single trenches suitable for specific wastes discharged direct from road tankers

Fence

Identification notice

Discharge pipe

Fence

Zig-zag pattern of trenching suitable for pumped discharge from reception tanks containing compatible wastes

Leachate monitoring points

Warning notices

Up to 100m

SECTION VIEW

Identification notice

OPEN TRENCH

Warning notices

COVERED TRENCH

Identification notice

Discharge pipe

Fences

Discharge pipe

Fence

Cover material

Fence

Household/ commercial waste

2 to 4m

Household/ commercial waste

Coarse fill eg. rubble, motor tyres

Liquid/sludge waste

Liquid/sludge waste

0.5 to 2m

0.5 to 2m

Codisposal of awkward wastes

7.113 *Dusty wastes* Occasionally wastes arise which are extremely fine and light. Unless they are handled and landfilled with due care and attention, such wastes can give rise to significant dust problems, both on the landfill and possibly outside its boundaries.

7.114 Whenever practicable, such wastes should be either packaged, for example, in biodegradable material or dampened sufficient to suppress dust and then deposited in trenches. The area around the trenches should be kept damp to prevent dusty material drying out and the trenches should be backfilled as soon as practicable. Where it is not practicable to moisten the waste beforehand, it should be fed by gravity using water sprays into the trench through a large diameter pipe. It will often be necessary for site workers to wear appropriate respiratory protective equipment when disposing of dusty wastes. Dusty wastes which present an additional acute hazard due to the presence of toxic materials should not be landfilled directly but should be processed to remove the hazard before disposal.

7.115 *Asbestos* The disposal of asbestos wastes is discussed in Waste Management Paper No 18 (as amended) and in Guidance Notes issued by the Health and Safety Executive. The landfilling of asbestos requires particular care. All fibrous and dusty asbestos wastes should be accepted at a landfill site only in robust plastic sacks or similar wrapping, as required by regulation. These should be sufficiently strong that they do not burst during loading, transport and unloading. Intact articles made from asbestos cement or bonded asbestos need not necessarily be bagged. All fibrous or dusty asbestos waste should be delivered to the site as separate consignments. On arrival, the delivery vehicle should be directed to the designated disposal area. The current practice is for fibrous or dusty asbestos to be deposited at the base of a working face or preferably into previously excavated trenches. The asbestos waste should be carefully deposited and then immediately covered. Vehicles should not be allowed to track through bags or drag them around. At least 0.5 m, of other suitable wastes must be immediately spread over bagged loose asbestos, but 0.2-0.25 m is sufficient for hard bonded asbestos, such as asbestos cement. In addition, the top, sides and face of the deposited asbestos should not be less than 0.5 m from the surface or flanks of the current working face. Asbestos wastes should not be deposited within the top 2 m of a site.

7.116 Procedures should be prepared for the eventuality that bags containing fibrous or dusty asbestos waste burst during handling or disposal at a site. The procedures should ensure that exposure of all personnel to airborne asbestos fibres is kept as low as possible and in any case does not exceed the relevant control limits. Such procedures should include:

— immediate damping down of any asbestos spillage;

— immediate covering of any spillage with a substantial layer of other waste in a manner which avoids the generation of airborne dust;

— where personnel are likely to be exposed to asbestos dust they should wear adequate respiratory protection of a type approved by the Health and Safety Executive;

— if personnel become contaminated they should change outer clothing, shower and dispose of contaminated clothing after suitable bagging;

— if a vehicle or equipment becomes contaminated it should be washed thoroughly and the washings disposed of in an acceptable manner.

7.117 *Odorous wastes* A number of industrial wastes including those from the intensive rearing of animals and by-product rendering and some chemicals manufacture, can give rise to odours. Odours emanating from landfill sites must be prevented generally and particularly when the site is located near to housing or industrial developments. For this reason, many sites are regarded as being unsuitable to receive odorous wastes. Adequate waste reception and handling facilities are essential to controlling obnoxious industrial wastes. Prior notification of delivery of such wastes is normally necessary and its reception and disposal should be undertaken only in suitable weather conditions. Various techniques are available to minimise odours from wastes being landfilled. These include below surface injection of liquids and the immediate covering of solids with materials possessing odour suppressing properties.

7.118 *Drummed wastes* As a general principle drummed waste is unacceptable at landfill sites. Large quantities of drums pose stability problems when landfilled since it is impossible to eliminate all void space between the drums and there is invariably ullage space in the drums. Drums eventually corrode away and release their contents into the body of a landfill site. Despite the precautions described below it is seldom possible to guarantee the contents of all drums. There is a risk therefore of the disposal into a landfill of substances that would be unacceptable. However, when, in exceptional circumstances and for sound technical reasons, the landfilling of waste packaged in drums is the best practicable environmental option, the total volume of such wastes landfilled should be carefully regulated. Practices used should prevent localised concentrations of drums occurring. Only full drums of particular types of solid wastes eg certain distillation residues, with softening points well above temperatures likely to be encountered in landfill sites, cast into drums, should be allowed.

7.119 Wastes in drums are frequently heterogeneous and are often difficult to identify. Careful and detailed labelling by the waste producer of the contents of each drum which accurately describes the contents, is essential.

Nevertheless, the landfill operator cannot be certain, without inspecting the contents of each drum, that the wastes he is accepting are as described. All drums should therefore be inspected on arrival at a landfill site to ensure that the contents conform to their description and labelling. To facilitate this, only lidded, open top drums should be used to contain the waste. Even with regular, consistent composition waste arisings, the operator should confirm with the producer that the composition of each consignment of waste has not been subject to significant changes.

7.120 Drums of wastes containing dangerous substances should be labelled in accordance with The Classification, Packaging and Labelling of Dangerous Substances Regulations 1984 (SI 1984 No 1244). In addition, it is recommended that labels detailing their contents should be affixed to each drum. Landfill operators should not therefore accept any drummed waste unless they are satisfied that they know by inspection or analysis the contents in each drum and that the acceptance of these materials is permitted by their disposal licence.

7.121 Inevitably, the contents of drums for disposal will not always be adequately identified; this is particularly the case when factories are being cleared out. A waste disposal contractor will therefore need to be provided with the results of chemical analysis of each drum of waste before he can decide how best to deal with such an arising. Good landfill practice demands that only those wastes whose composition is known to be compatible with the landfill environment, should be landfilled. Chemical treatment or incineration will be the appropriate method of disposal for some wastes. In any case the landfill operator will need to be satisfied that the contents of drums are appropriate for landfill disposal based on the results obtained from sampling and analysis. Sampling should be carried out at the point of arising to satisfy paragraph 7.120 above.

7.122 Before any drum is sampled, attempts should be made to establish at least the range of substances which might be in each drum. During opening and subsequent sampling, precautions taken and the protective equipment worn should be appropriate to the risk posed by the most hazardous substance likely to be found. Once the contents have been identified each drum should be labelled accordingly. Where appropriate, consignment notes should be obtained if the material is 'Special Waste' under the Control of Pollution Special Waste Regulations (see paragraph 2.68) and the appropriate procedures followed.

7.123 At landfill sites accepting drummed waste a designated reception and storage area is required. This should be of a suitable size to allow lorries to manoeuvre and park safely. The storage area should afford adequate security and be of a size sufficient to hold at least one days quantity of drums suitable for landfill disposal. As far as practicable, it is recommended that inspection and sampling of all drums takes place with the drums still on the vehicle. Wastes found to be unsuitable for landfilling should be returned to the waste producer for more appropriate disposal elsewhere eg at a waste treatment facility. Only when the inspection has proved that the wastes are acceptable should the drums be unloaded. Unloading should be supervised and by the use of appropriate handling equipment precautions taken to avoid accidental damage to the drums and spillage of their contents.

7.124 The disposal of all drummed wastes found to be acceptable should be undertaken without delay. Drums should be spaced out across the base of the working face at intervals of not less than 0.5 m and immediately completely surrounded and covered with compatible waste or other cover materials. Some drummed waste, eg solid pesticide formulations should be emptied out and the contents spread as widely as possible in admixture with biodegradable waste within the landfill.

7.125 The disposal of drums containing liquid wastes presents many problems. Such drums should be emptied and their contents disposed of in accordance with good landfill practices such as trenching or lagooning. Normally, decanting and bulking up of drummed liquids should always be undertaken at their point of arising. However, it is recognised that in isolated instances it will not be practicable or safe to do so. In such cases, the drums should not be landfilled but their contents decanted at the landfill site into trenches or lagoons designated for the particular type of waste. Again, suitable precautions must be taken at all stages to ensure the safety of operatives. It must also be ensured that only compatible wastes are deposited and that liquid loading rates are not exceeded. The re-use of drums that have contained liquid waste as containers for other materials can be economically viable, but does present health and safety risks unless they can be effectively decontaminated. Reuse of drums that have previously contained waste is not a practice that can be recommended; generally drums should be disposed of separately after crushing or they can be sold as scrap metal after having first been thoroughly cleaned and crushed.

7.126 *Disposal of tyres* The landfilling of tyres in large quantities can lead to considerable difficulties especially from fires, stability and settlement and leachate quality. Consequently, tyres should not be landfilled unless they have previously been finely shredded or are deposited in extremely small numbers. It is recommended that the quantities of shredded tyres landfilled should be strictly regulated and should not exceed 5% by volume of the total input of solid wastes. Intact tyres may be used in limited quantities for the construction of injection wells (see paragraph 7.107).

7.127 Within a landfill conditions are such that a wide range of organic compounds can be leached from tyres. Leachate migration in such circumstances can cause severe pollution problems. Unfortunately, there have also been a number of fires involving tyres at landfill sites. These can be difficult to contain and extinguish. Fires involving large quantities of tyres can result in the pyrolytic breakdown of rubber, with the production of complex mixtures of highly polluting products.

7.128 At landfills where significant quantities of tyre fragments (up to 5%) are deposited they should be placed in specially constructed cells or trenches not more than 2 m deep. On completion, the cells should be covered with at least 1 m of incombustible material. Special fire-fighting procedures should be laid down and incorporated into a contingency plan agreed with the local fire-brigade.

Cover requirements

7.129 Cover requirements and materials are discussed in Chapter 5 (paragraph 5.28 et seq). An additional consideration with difficult wastes is the need to ensure that any cover materials used are compatible with the wastes.

Equipment

7.130 Equipment requirements are considered, in general, in Chapter 5. The disposal of difficult wastes may require additional specialised equipment to be available.

7.131 Mobile pumps which are capable of handling a range of liquids and sludges will be required. Pumps will also be required for spraying and irrigating liquid wastes. For the disposal of solid difficult wastes special equipment is not normally required except for handling drums. It is important to ensure that any equipment that becomes contaminated does not lead to waste being spread throughout the site. Equipment therefore should be regularly cleaned. An efficient wheel cleaning facility for vehicles is essential (see paragraph 4.18 et seq).

7.132 The provision of washing and first aid facilities is discussed at paragraph 4.8, the provision of protective clothing is also discussed at paragraphs 7.148-149. Additionally, mess and changeroom facilities must include a shower and preferably an emergency water drench in addition to any drench showers provided in the working area.

Monitoring of wastes

7.133 Monitoring is discussed in various sections of this report (see paragraphs 3.66 et seq, 3.70, 5.9 and 5.76). At sites accepting difficult wastes it is particularly important to analyse leachate as well as the incoming wastes.

Analysis of leachate

7.134 Leachate at codisposal sites should be monitored regularly to ensure that attenuation mechanisms are not being overloaded and that attenuation is taking place as predicted. For sites where leachate treatment is undertaken or where leachate is being discharged to sewer, monitoring is required to ensure that effluent quality is maintained within the design parameters of the plant and meets the consent conditions for its discharge.

Analysis of incoming waste

7.135 Inspection and analysis of incoming waste is necessary to verify the waste producer's description of the waste, to ensure the health and safety of personnel handling the waste, its compliance with disposal licence requirements, and to confirm that the waste is suitable for the method of disposal chosen. Detailed analysis on-site is not normally necessary. Nevertheless, regular monitoring and sampling of wastes received should be carried out and if necessary samples sent for more detailed analysis to a central laboratory.

7.136 The analytical facilities required at a landfill site accepting difficult wastes will depend on the nature and quantity of incoming wastes. Small sites which can accept only a limited range of difficult wastes may need to provide only basic analytical equipment and rely on external consulting analysts for any detailed information required. Common tests include:

(a) appearance

(b) odour

(c) pH

(d) flammability

(e) specific gravity.

7.137 Equipment for a laboratory serving a landfill site accepting difficult wastes is suggested in Table 7.137.

Table 7.137 Site laboratory equipment

Equipment	Parameter
1. Sampling devices for liquid/sludge wastes	
2. pH papers, or pH meter	pH
3. Titration equipment	acid/alkali strength
4. Absorbent paper and matches	flammability
5. Glassware	appearance odour
6. Dropper bottles of selected reagents	reactivity
7. Bellows and a selection of gas detection tubes	volatility and gas type
8. Filtration equipment	
9. Hydrometer	specific gravity

Manning levels

7.138 Manning levels and the required qualifications of staff will be determined by the quantity and type of waste handled. Monodisposal lagoons and some multidisposal in-house landfills may not need to be attended, except when disposal is taking place from vehicles rather than pipelines, for regular inspections and security checks. However, codisposal landfills will inevitably require additional manpower in comparison with sites accepting only household or similar wastes. No precise advice can be given, but the following may be regarded as key personnel in the satisfactory operation of a landfill taking difficult wastes.

7.139 *Site manager* A landfill site accepting significant quantities of difficult wastes must have a permanent site manager responsible for ensuring that all waste disposal operations are carried out in accordance with the operational plan and that all regulations are complied with. He will also have to make decision on whether to accept wastes, suitable loading rates, appropriate handling procedures for differing types of waste and the action to be taken if unplanned events occur. Site managers therefore, should have considerable experience in all aspects of landfill operations.

7.140 *Site foreman or chargehand* He will be responsible for the day-to-day running of the site. His primary duty must be to ensure that the site is run in accordance with the procedures laid down in the Operations Manual.

7.141 *Face-marshall* At sites where the volume of incoming traffic is high and where a wide range of wastes are delivered, a face marshall will normally be required. He will be responsible for directing vehicles to the appropriate operational area and ensure that their loads are discharged in a satisfactory manner.

7.142 *Gate-marshall or weighbridge operator* A record of all in-coming loads is essential. Where wastes are consigned under the 'Special Waste' regulations he will be responsible for checking that the consignment note which accompanies such waste agrees with the load and is expected.

7.143 *Machine operators* The number of operatives required will depend on the quantity of equipment used on the site at any one time.

Training

7.144 All operatives should be trained to a standard which will satisfy the requirements of Health and Safety legislation (see Chapter 2) particularly in the safe operation of equipment. They should be aware of the working plan for the site and the standards of operation to be carried out. The nature and types of materials being handled, and the difficulties involved in ensuring continuous supervision of all employees makes it essential that staff receive a particularly high standard of training.

Health and safety

7.145 A major responsibility when running a landfill site which accepts difficult wastes is the protection of site operatives and delivery personnel from risks to their health and safety. An additional consideration is the safety of visitors and the general public who may live near the site. The statutory basis of these duties is outlined in Chapter 2.

7.146 The effective and continuous protection of landfill site operatives from risks may be difficult to achieve. Such difficulties however, cannot be used as an excuse to ignore the issue. The standards of safety, training and supervision set should be commensurate with the risks involved and should be monitored regularly to ensure that they are effective.

7.147 Many hazards which exist at landfill sites, are common to other areas of employment, such as the extensive use of mobile plant and vehicles. It is essential therefore that those responsible for safety matters at a landfill site recognise and deal both with the normal hazards and those which are unique to landfill operations. These include:

(i) the hazards which may result from consignors failing to declare the true and precise nature of a consignment of waste, and

(ii) the hazards which may result from the mixing of incompatible wastes — this hazard may be aggravated by a lack of knowledge of the precise composition of the waste eg Gypsum in admixture with household waste can generate hydrogen sulphide and under similar conditions arsine can be produced by arsenic compounds.

7.148 Employees at landfill sites work in all weather conditions and will need to be provided with suitable windproof wet weather clothing. Weather conditions can also make the handling of some wastes more difficult, for example the handling of dusty wastes will require special care in dry, windy conditions, while heavy rain may increase the difficulty of spreading sludge. The disposal of even small quantitites of solvent bearing wastes (see paragraph 7.82) may be hazardous in hot, still weather.

7.149 Most hazards can be minimised by adopting working practices to lessen the hazard. In most instances the provision of appropriate handling equipment and suitable protective clothing will be necessary. The type of protective clothing used should be carefully selected to ensure not only that it is adequate for the task to be performed but that its design ensures employees' safety and is comfortable and practicable when worn under the arduous conditions which can prevail on landfill sites.

Operational procedures

7.150   All waste should be checked on delivery to the site and if found to be acceptable should be directed to the appropriate disposal area. Where a reception sump for liquid wastes is not available, a site operative should accompany the tanker to or receive it at the designated disposal area. Vehicles carrying difficult solid wastes or sludges should be directed to either the base of the working face, or to some other designated area of the site. In both cases it will be necessary to ensure that the site roads are adequate for the traffic flow. Under bad site conditions (particularly in winter) it may be necessary to transfer loads to on-site vehicles for carriage to the disposal area.

7.151   Machine drivers and face marshalls should be made aware of the importance of inspecting loads as they are being deposited to ensure that unacceptable wastes are segregated for disposal elsewhere.

7.152   'Totting' should not be permitted on landfill sites taking difficult wastes. Lone working as far as practicable should not be allowed on such sites and where this is not possible some means of regularly checking the well-being of all personnel on a landfill site should be practiced.

7.153   Fires at landfill sites accepting difficult wastes can present additional special problems due to the noxious and possibly toxic fumes and particulates which may be generated. Operators manning equipment to bring fires under control should wear self-contained breathing apparatus. If dense fumes are being evolved, even when the cab of a site vehicle is air conditioned, such apparatus should be worn.

First aid arrangements

7.154   First aid facilities have been discussed in Chapter 5. There is always a possibility that employees could be could be splashed by irritant or corrosive materials. Consideration should be given therefore to the provision of equipment such as eye wash bottles or other means of irrigating the eyes. Contamination of large areas of the body is also possible and the provision of emergency showering or drenching facilities with a continuous supply of clean water near to the working area as well as centrally should be considered (see also paragraph 7.132). Good communications are essential on large landfill sites accepting difficult wastes. The usefulness of radio telephones for rapid site communications has been proven.

The restoration of landfills that have taken difficult wastes

Introduction

7.155   Site restoration is discussed in detailed in Chapter 6, but the codisposal of difficult wastes raises additional considerations, in particular the prevention of potentially hazardous materials reaching the surface of a restored landfill as well as their release to the environment in an uncontrolled manner.

Transport mechanisms

7.156   Various transport processes may result in the migration of toxins into the surface cover or surroundings of a landfill. If the substance or its precursor is not already in a mobile form then this will usually involve solubilisation or volatilization of the substance itself, or chemical or biological reactions forming soluble or gaseous products. Figure 7.156 shows the possible pathways by which this may occur.

Transport to soil

7.157   Gases and vapours in the landfilled waste may migrate through the soil cover or move laterally through porous layers or fissures to emerge some distance from the site. Substances dissolved in leachate may emerge from a landfill in seepages. Under certain climatic conditions, upward movement of leachate by capillary action can occur and result in contamination of surface soil.

7.158   Once a substance has migrated away from a landfill site it possibly may be broken down by chemical or biological reactions in the soil. Alternatively it may form insoluble compounds, or become firmly bound to soil, or organic matter. Alternatively it may remain available for uptake by plants and be ingested subsequently by animals.

Transport to vegetation

7.159   Root penetration into landfilled waste is the most direct pathway by which vegetation may be adversely affected. Some wastes will be phytotoxic to plant life. If the waste is not phytotoxic to plants it may by various biological processes be translocated through the plant to accumulate in their stems, leaves or seedpods. The possibility of plant damage by asphyxiating vapours or gases in the root zone also exists.

Ingestion by animals

7.160   Animals, including man, can be exposed to toxic substances deposited on landfill sites unless the latter are properly restored and the wastes thus isolated. Likewise vegetation can also become contaminated. The ingestion of soils is inevitable in grazing animals. Sheep have been reported to ingest up to 40% soil while grazing. The 'pica syndrome' in children could also be an important consideration if the afteruse of an improperly restored landfill allowed public access. The possibility of exposure to toxic substances via the consumption of crops grown on contaminated land and the consumption of food-stuffs from animals grazed on contaminated areas is an important consideration.

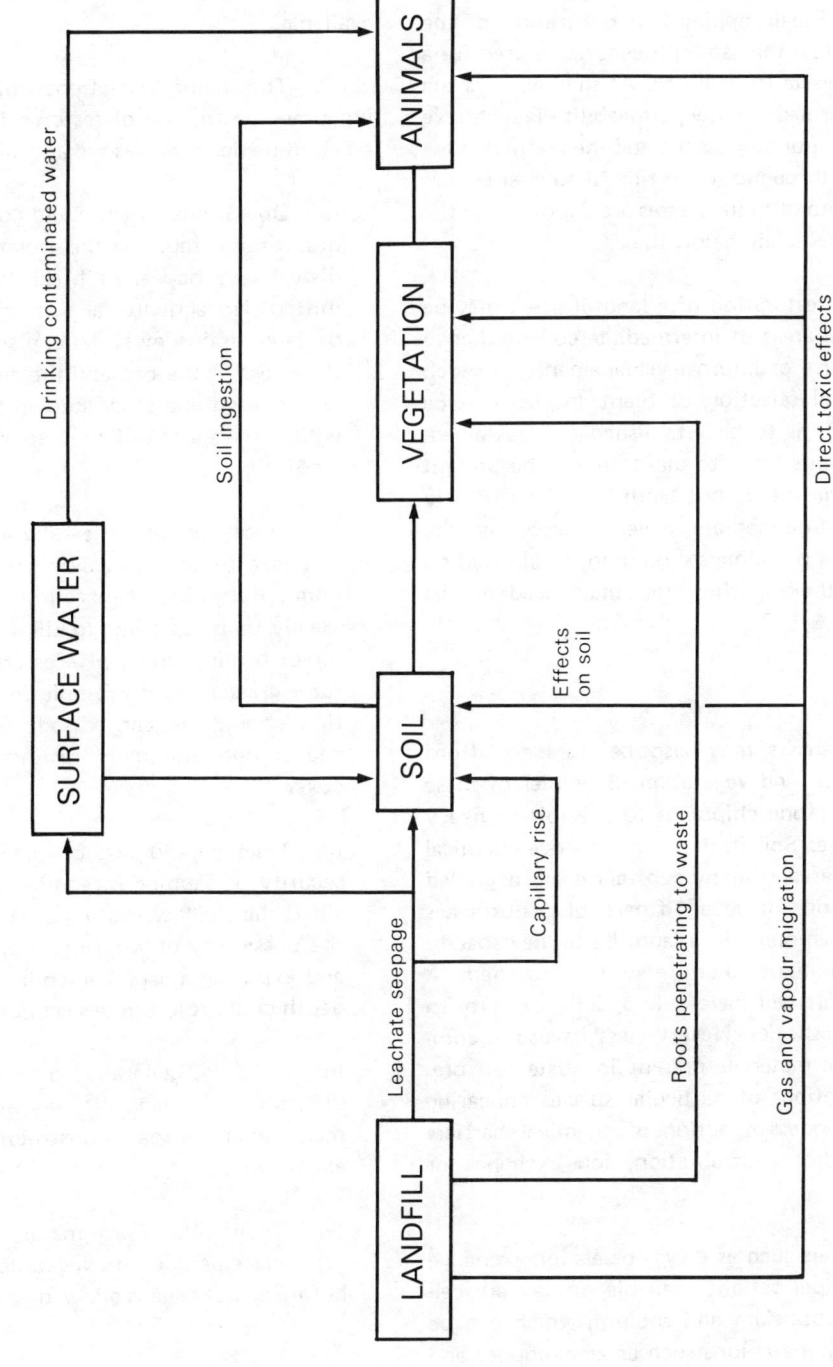

Figure 7.156   POSSIBLE MIGRATION OF TOXINS TO SOILS, VEGETATION AND ANIMALS AT RESTORED LANDFILL SITES

Protecting animals and vegetation from the toxic effects of difficult wastes

7.161 The restoration plan for a landfill site (see Chapter 3) should have considered the means by which difficult wastes are deposited within the landfill site. For example at deep sites difficult wastes should only be deposited at a considerable depth below the surface and covered with an intermediate low permeability clay cap. The final lift of waste should be limited to relatively inert household and commercial waste. Final capping and restoration of the site should then follow the same procedures as used for a normal household waste landfill site. At shallow sites the provision of an intermediate low permeability layer above difficult wastes may not be possible and the waste deposit will be continuous throughout the fill. At such sites it is recommended that no difficult wastes are deposited in the final 2 m of fill immediately below the cap.

7.162 Before final restoration of a landfill site it may be desirable to vegetate areas of intermediate cover either to stabilise the land form or improve visual amenity. In such instances the careful selection of plant species will be required if plant life is to be established and sustained. Not only contaminants toxic to plant life can be present in the cover material used, but landfill gas can usually permeate through intermediate cover. Access by the public or the grazing of animals should not be allowed on such areas. Nevertheless, the area may need to be husbanded.

Chemical barriers

7.163 Chemical barriers may also be used to afford protection to animals and vegetation. Examples of these are the use of limestone chippings to precipitate heavy metals from leachate. Soil itself is an effective chemical and biological barrier and many substances are degraded or immobilised in soil. However, if particular substances are present in large amounts in a landfill site the capacity of the site to attenuate them may be exceeded. A chemical barrier itself will have only a finite capacity to attenuate particular species. Nevertheless its use in combination with other materials present in waste can prevent high concentrations of particular species appearing in leachate. The modes of action of chemical barriers usually involves either neutralisation, ion exchange or precipitation.

7.164 Some materials such as clays possess ion exchange capacity. The principal cations available in clay are calcium, magnesium, potassium and sodium, which can be exchanged for heavy metal ions such as zinc, copper and nickel. The cation exchange capacity varies for individual clay minerals and only clays with at least moderate cation exchange capacities should be used. Most metals are precipitated as insoluble carbonates and phosphates, thus ground limestone and phosphate fertilisers are efficient at precipitating metal ions from solution.

Limitations on use of restored difficult waste landfills

7.165 The need for longer term security of landfill sites and the avoidance of future damage to restored land, may necessitate limitations on their afteruse. It is important that such limitations are clearly recorded and that present and future owners, local authorities and other persons with interests in the site are fully aware that the land is a restored landfill site. Records of the operation, restoration and aftercare undertaken at the site should be readily available.

7.166 The major restrictions which may apply for varying times on the use of restored landfill sites which have received difficult wastes are as follows:

(a) Operations which could compromise the effectiveness of an impermeable isolating cap or in any way disturb any barrier or break layer should not be permitted. No agricultural operations such as installation of land drains must take place within the cap. The thickness of the cap and the depth of soil cover should be recorded and made known to all persons concerned with management of the site after it has ceased to be a landfill.

(b) Once vegetation is fully established, physical disturbance of the soil cover should be kept to a minimum. Reseeding of grazed or harvested swards should ideally be by spraying to kill off the sward followed by direct drilling into it. Necessary cultivation of agricultural grassland for example to relieve surface compaction should be carried out using equipment which causes only minimum disturbance of the vegetation cover.

(c) Trees should not be planted on any site where security of capping is required. Unless mounds are provided the shallow rooting depths available can result in the possibility of windthrow disturbing and lifting soil and exposing a cap. The ability of trees to dry soils at depth could result in desiccation of a clay cap.

(d) Building, generally, should not be permitted on difficult waste landfill sites nor should any development such as road construction which involves re-excavation.

(e) The effects of any major change in use of the site or management of its vegetation should be evaluated before any change is allowed.

PAPERS IN THE WASTE MANAGEMENT SERIES (see also Annex 3)

WMP No	Title
1	Reclamation, Treatment and Disposal of Wastes — An Evaluation of Options, HMSO 1976
6	Polychlorinated Biphenyl (PCB) Wastes, HMSO 1976
7	Mineral Oil Wastes, HMSO 1976
8	Heat Treatment Cyanide Wastes, HMSO 1976
9	Halogenated Hydrocarbon Solvent Wastes from Cleaning Processes, HMSO 1976
11	Metal Finishing Wastes, HMSO 1976
12	Mercury Bearing Wastes, HMSO 1977
13	Tarry and Distillation Wastes and Other Chemical Based Residue, HMSO 1977
14	Solvent Wastes (excluding Halogenated Hydrocarbons), HMSO 1977
15	Halogenated Organic Wastes, HMSO 1978
16	Wood Preserving Wastes, HMSO 1980
17	Wastes from Tanning, Leather Dressing and Fellmongering, HMSO 1978
18	Asbestos Wastes, HMSO 1979
19	Wastes from the Manufacture of Pharmaceuticals, Toiletries and Cosmetics, HMSO 1978
20	Arsenic Bearing Wastes, HMSO 1980
21	Pesticide Wastes, HMSO 1980
23	Special Wastes: A technical memorandum providing guidance on their definition, HMSO 1981
24	Cadmium-Bearing Wastes, HMSO 1984
25	Clinical Wastes, HMSO 1983

EC DIRECTIVE ON THE PROTECTION OF GROUNDWATER AGAINST POLLUTION CAUSED BY CERTAIN DANGEROUS SUBSTANCES (see paragraphs 7.62 and 3.5)

List I: Substances to be prevented from entering groundwaters

1. Organohalogen compounds and substances which may form such compounds in the aquatic environment.

2. Organophosphorus compounds.

3. Organotin compounds.

4. Substances which possess carginogenic, mutagenic or teratogenic properties in or via the aquatic environment.

5. Mercury and its compounds.

6. Cadmium and its compounds.

7. Mineral oils and hydrocarbons.

8. Cyanides.

List II: Substances to be limited so as to avoid pollution

1. The following metalloids and metals and their compounds:

    1. Zinc
    2. Copper
    3. Nickel
    4. Chromium
    5. Lead
    6. Selenium
    7. Arsenic
    8. Antimony
    9. Molybdenum
    10. Titanium
    11. Tin
    12. Barium
    13. Beryllium
    14. Boron
    15. Uranium
    16. Vanadium
    17. Cobalt
    18. Thallium
    19. Tellurium
    20. Silver

3. Substances which have a deleterious effect on the taste and/or odour of groundwater, and compounds liable to cause the formation of such substances in such water and to render it unfit for human consumption.

4. Toxic or persistent organic compounds of silicon, and substances which may cause the formation of such compounds in water, excluding those which are biologically harmless or are rapidly converted in water into harmless substances.

5. Inorganic compounds of phosphorus and elemental phosphorus.

6. Fluorides.

7. Ammonia and nitrites.

CLASSIFICATION OF DIFFICULT WASTES

Type of waste	Groups and sub-groups	Group code	Sub-group code
A Inorganic acids	Hydrochloric acid	A10	
	Sulphuric acid	A20	
	Nitric acid	A30	
	Chromic acid	A40	
	Phosphoric acid	A50	
	Hydrofluoric acid	A60	
B Organic acids and related compounds	Aliphatic acids eg formic, acetic and oxalic acids		B11
	Aromatic acids eg benzoic, phthalic acids		B12
	Acid anhydrides eg acetic, phthalic anhydrides		B13
	Acid chlorides eg acetyl, benzoyl chlorides		B14
	Sulphonic acids		B15
C Alkalis	Alkali metal oxides and hydroxides, calcium oxide, proprietary alkaline cleaners	C10	
	Sodium and/or potassium hydroxides or oxides		C11
	Calcium oxides		C12
	Ammonia	C20	
	Calcium hydroxide		C91
	Sodium and/or potassium carbonates		C92
D Toxic metal compounds	Cadmium	D10	
	Mercury	D20	
	Lead	D30	
	Arsenic	D40	
	Copper		D91
	Zinc		D92
	Barium (water soluble forms)		D93
	Thallium		D94
	Nickel		D95

Type of waste	Groups and sub-groups	Group code	Sub-group code
	Vanadium		D96
	Silver		D97
E Non-toxic metal compounds	Ammonium salts		E91
	Titanium		E92
F Metals (elemental)	Alkali, alkaline earth and other hazardous metals	F10	
G Metal oxides	Hazardous oxides	G10	
	Cadmium oxide		G11
	Beryllium oxide		G12
H Inorganic compounds	Cyanides	H10	
	Sodium and potassium cyanides		H11
	Soluble complex cyanides		H12
	Ferro and ferricyanides		H13
	Sulphides, selenides, tellurides and arsenides		H21
	Oxidizing compounds	H30	
	Hypochlorites and chlorites		H31
	Chlorates, perchlorates, bromates, iodates, periodates, persulphates and permanganates		H32
	Peroxides		H33
	Chromates		H41
	Fluorides, silicofluorides, borofluorides		H42
	Arsenates and arsenites		H43
	Carbides and acetylides		H91
	Borates		H92
	Nitrites		H93
	Nitrates		H94
J Other inorganic materials	Asbestos	J10	
	Slag including boiler and flue cleanings	J20	
	Mineral processing wastes	J30	
	Silt and dredgings	J40	
	Water (contaminated)	J50	

Type of waste	Groups and sub-groups	Group code	Sub-group code
K Organic compounds	Hydrocarbons (not included in M)	K10	
	Aliphatic hydrocarbons		K11
	Aromatic hydrocarbons		K12
	Phenols, analogues and derivatives	K20	
	Chlorinated phenols and analogues		K21
	Peroxides	K30	
	Halogenated cleaning compounds	K40	
	Trichloroethylene		K41
	Perchloroethylene		K42
	Trichlorethane		K43
	Trichlorotrifluoroethane		K44
	Halogenated compounds excluding cleaning compounds	K50	
	PCBs and analogues		K51
	Organo metallics	K60	
	Tetra ethylead		K61
	Tetra methylead		K62
	Nitrogen, sulphur or phosphorus-containing compounds	K70	
	Amines and amides		K71
	Nitro compounds		K72
	Nitriles		K73
	Isocyanates		K74
	Other organo nitrogen compounds		K75
	Organophosphorous compounds		K76
	Organosulphur compounds		K77
	Oxygen containing organic compounds	K80	
	Esters		K81
	Ethers		K82
	Aldehydes and ketones		K83
	Alcohols		K84
	Chelating compounds		K91
	Phthalates		K92

Type of waste	Groups and sub-groups	Group code	Sub-group code
L Polymetic materials and precursors	Precursors, monomers and products of incomplete polymerization	L10	
	Epoxy resins (not finished products)		L11
	Polyester resins (not finished products)		L12
	Phenol-formaldehyde resins (not finished products)		L13
	Finished products and manufacturing scrap	L20	
	Polyurethane		L22
	Scrap rubber (including tyres)	L30	
	Latex, latex and rubber solutions and suspensions	L40	
	Synthetic adhesive wastes	L50	
	Ion-exchange resin wastes	L60	
M Fuel, oils and greases	Mineral oils	M10	
	Kerosene and derv	M20	
	Fuel oil	M30	
	Vegetable and other oils	M40	
	Oil/water mixtures	M50	
	Fats, waxes and greases	M60	
N Fine chemicals and biocides	Pharmaceutical and cosmetic products	N10	
	Pharmaceutical products in retail containers		N11
	Pharmaceutical products in bulk and production containers	N13	
	Biocides	N20	
	Pesticides		N21
	Herbicides		N22
	Fungicides		N23
P Miscellaneous chemical waste	Organics identified by trade names* only		P31
	Inorganics identified by trade names* only		P32
Q Filter materials, treatment sludge and contaminated rubbish	Used filter materials eg kieselguhr, carbon, filter cloths	Q10	
	Contaminated rubbish (including bags and sacks)	Q20	
	Empty used containers	Q30	
	Industrial effluent treatment sludges	Q40	

* Where trade names are used the source of the material should be specified.

Type of waste	Groups and sub-groups	Group code	Sub-group code
R Interceptor wastes, tars, paint, dyes and pitments	Tank cleaning sludge (note K60 for lead contents)	R10	
	Interceptor pit wastes (note M10-M30 for oil content)	R20	
	Printing industry wastes (ink manufacture and use)	R30	
	Dyestuffs waste	R40	
	Distillation residues	R50	
	Acid tars	R60	
	Tar, pitch, bitumen and asphalts	R70	
	Paint waste (manufacture and use)	R80	
Miscellaneous wastes	Tannery and fellmongers waste	S10	
	Tannery waste		S11
	Fellmongers waste		S12
	Cellulose wastes (natural and synthetic)	S20	
	Waste treated timber	S30	
	Soap and detergents	S50	
	Soap		S51
	Detergents		S52
Animal and food wastes	Animal processing wastes	T10	
	Carcasses and flesh		T11
	Blood, fat, grease etc		T12
	Excrement		T13
	Food processing wastes (including starch)	T20	
	Glue wastes	T30	

MEMBERSHIP OF THE LANDFILL PRACTICES REVIEW GROUP
(see paragraph 1.12)

CHAIRMEN

Mr J Bentley   Department of the Environment
Mr B Gulley   Department of the Environment

CHAIRMEN OF SUB-GROUPS

Mr N Harrison   National Association of Waste Disposal Contractors
Mr A Higginson   Institute of Wastes Management

SECRETARIAT

Mr S Carlyle   Department of the Environment
Mr J Grayson   Department of the Environment
Mr A Parker   Harwell Laboratory
Mr K Pearce   Department of the Environment
Dr G Rae   Department of the Environment
Mr P Young   Harwell Laboratory

MEMBERSHIP

Mr R Aspinwall   Aspinwall & Company
Dr C Barber   Water Research Centre
Mr N Beard   Ministry of Agriculture, Fisheries and Food (from July 1983)
Mr M Beckett   Department of the Environment
Mr J Bentley   Department of the Environment
Mr C Biddle   Packington Estate Enterprises Ltd
Mr J Boldon   West Midlands Metropolitan County Council
Mr C Burford   Lancashire County Council (representing Waste Disposal Engineers Association)
Mr R Caisley   Derbyshire County Council
Mr D Campbell   Harwell Laboratory
Mr A Cheyney   London Brick Landfill
Mr A Constantine   Suffolk County Council
Dr D Dixon   Scottish Development Department
Mr J Dobson   Tyne & Wear Metropolitan County Council
Mr P Edwards   ICI Mond Division (representing Chemical Industries Association)
Mr L Graham   Delyn Borough Council (until April 1983) (representing Association of District Councils)
Mr K Guiver   Southern Water Authority
Mr B Gulley   Department of the Environment
Mr R Harris   Severn-Trent Water Authority
Mr R Haythornthwaite   Ministry of Agriculture, Fisheries and Food (until July 1981)

Mr J Hewitt	Tyne & Wear Metropolitan County Council
Mrs T Hillman	West Midlands Metropolitan County Council
Mr B Hurley	Cory Sand & Ballast
Dr R Keen	Bristol Polytechnic
Mr L Knott	Cheshire County Council (representing Association of County Councils)
Mr K Knox	Cleanaway Ltd
Mr J Lucas	Aspinwall & Company
Mr D Mills	Department of the Environment (until May 1983)
Mr A Milroy	Shanks & McEwan Ltd
Mr D Morris	Health and Safety Executive
Mr J Newton	Amey Roadstone Ltd
Mr R Osmond	Department of the Environment
Mr C Palmer	Suffolk County Council
Mr P Parsons	Department of the Environment
Mr A Pearce	Department of the Environment
Mr D Platten	Hargreaves Clearwaste Ltd
Mr B Price	Friends of the Earth
Mr T Roberts	Montgomery District Council (from July 1983)
Mr T Robson	Hargreaves Clearwaste Ltd
Mr J Searle	Consultant
Mr A Smith	North West Water Authority
Mr P Sullivan	Health and Safety Executive
Mr J Tankard	Health and Safety Executive
Mr M Tassel	Cambridgeshire County Council
Mr G Taylor	Welsh Office (until June 1983)
Mr C Tunaley	South Yorkshire Metropolitan County Council (representing Association of Metropolitan Authorities)
Mrs A Ward	Department of the Environment
Mr D Wilkinson	Cleanaway Ltd
Mr G Williams	British Geological Survey
Mr J Wilson	Scottish Development Department
Mr C Young	Water Research Centre

Advice was also received from:

Mr B Chapman	Department of the Environment
Mr L Clark	Water Research Centre
Mr T Day	Department of the Environment
Mr J Davies	Aspinwall & Company
Miss L Heasman	Harwell Laboratory
Mr J Lucas	Department of the Environment
Mr P Maris	Water Research Centre
Mrs S Proudman	Department of the Environment
Mr D Robinson-Todd	Stablex Ltd
Mr P Rushbrook	Harwell Laboratory
Miss S Sharpey	Department of the Environment
Dr C Stevens	Harwell Laboratory
Mr A Stuart	British Geological Survey
Mr N Walker	Cleanaway Ltd
Mr D Wright	Stablex Ltd

In addition a Synopsis of the LPRG Report was prepared for the Department by Mr M J T Mayer of Environmental Data Services Ltd. Mr L Rutterford, lately of the Department, advised on restructuring the Report into a Waste Management Paper.

THE CONTROL OF LANDFILL GAS

FOREWORD

At a landfill gas workshop organised by the Harwell Laboratory and the British Anaerobic and Biomass Association Ltd (BABA) at St Catherine's College, Oxford in September 1983, the need for guidance on the control and use of landfill gas was recognised. The Department of the Environment supported this proposal and set up the Landfill Gas Standards Committee comprising parties from all sectors with an interest in landfill gas. It quickly became apparent that the work of this committee was complementary to that of the Landfill Practices Review Group (LPRG). The committee was thus charged with the preparation of a report for inclusion in the main report of the LPRG. This annex constitutes that report.

The terms of reference for the Landfill Gas Standards Committee were:

(1) To investigate the health and safety aspects of controlled extraction of landfill gas, collection, venting or flaring and distribution to consumers for commercial use.

(2) To check existing regulations covering mines and quarries, the gas industry and any others which may be relevant, in order to extract and bring together those regulations deemed to be applicable.

(3) To prepare a report for submission to the Landfill Practices Review Group for publication.

The membership of the committee was as follows:

Chairman

Mr A J Smith, Department of the Environment

Secretary

Mr D Campbell, Harwell Laboratory

Members

Mr J Bentley, Department of the Environment
Mr W Bowers, Department of Energy
Mr S Carlyle, Department of the Environment
Mr J Clegg, Health and Safety Executive
Dr J Coombs, British Anaerobic and Biomass
   Association Ltd
Mr W L Hall, NCB (Coal Products) Ltd

Mr M Humphries, Kent County Council
Mr R J Lucas, Department of Energy
Mr A J Marchant, Greater London Council
Mr H Moss, National Association of Waste Disposal
   Contractors
Mr C Palmer, Suffolk County Council
Mr P Parsons, Department of the Environment
Mr K Pearce, Department of the Environment
Dr L Penney, Harwell Laboratory
Dr G Rae, Department of the Environment
Dr J Rees, BioTechnica Ltd
Dr K Richards, Department of Energy (ETSU)
Mr J Tankard, Health and Safety Executive
   (Mines and Quarries)
Mr D Taylor, Aveley Methane Ltd

1. INTRODUCTION

1.1 The production of landfill gas is an inevitable consequence of the disposal and subsequent decomposition of wastes containing organic materials to landfill. Such wastes will be predominantly domestic or commercial in origin but may include some industrial waste materials. Ever stricter environmental standards are being required by both practitioners within the waste disposal industry and by the general public. Thus attention has recently been focussed on the fate of the end products of refuse fermentation, including landfill gas, and in some cases on the potential recovery of this gas as a fuel source for use in industry.

1.2 The section on landfill gas (Chapter III, paragraph 3.71 et seq in the main report), discusses the various waste decomposition processes resulting in gas production, the potential problems of gas migration, odour nuisance and the opportunity for commercial recovery of gas. It should be referred to in conjunction with this Annex which considers the various technical aspects of implementing a gas control or recovery scheme to satisfy existing health and safety legislation and standards and to comply with the demands of other statutory bodies (planning, site licensing, Energy and Pipeline Acts). It is intended that this Annex will provide the framework for the rapid implementation of any proposed project. This is particularly important where an urgent decision is required for a gas migration control scheme.

1.3 Landfill sites vary widely in type from small, shallow sand and gravel excavations to large, deep quarries (hardrock, limestone, chalk, clay). Other variables include waste type, input rates and methods of compaction. Many of these variables will affect the timescale for the introduction of a gas migration control or gas utilization scheme, particularly if the site is still operational, and may also affect the extent and scale of the installed plant. Where commercial utilization or migration control is envisaged there are several aspects of site design, operation and restoration which are important. For example, a landfill scheme involving progressive restoration is desirable to enable gas control wells or trenches to be installed before significant gas production commences. Such a method of operation allows for control of water input to the waste which is an important parameter in gas production. Good compaction of waste encourages anaerobic fermentation by restricting air ingress.

1.4 The need to control or exploit landfill gas has required the application of various existing technologies, such as drilling techniques, plant, equipment and gas pipeline technologies. At a recent symposium on landfill gas it was agreed that certain standards are required in the interests of health and safety. Whilst some existing legislation is relevant to landfill gas there are some omissions or difficulties of interpretation, for example, where the legislation has been primarily aimed at other sectors of the gas industry, which could lead to variations in practice.

1.5 The following sections have been structured to follow the probable course of events at a site investigation. Although this document does not consider the economic viability of the commercial utilisation of landfill gas, reference to this aspect has been made where it impinges on such factors as safety, plant and equipment specifications and the requirements for pipelines to deliver gas to the user.

2. LANDFILL GAS COMPOSITION, HAZARDS AND RISK AVOIDANCE

Composition

2.1 The gas produced during the anaerobic decomposition of organic material present in landfilled wastes comprises largely methane and carbon dioxide (in the ratio of about 3:2), with relatively small percentages of oxygen and nitrogen. Hydrogen is often observed at up to about 15% in the early stages of refuse fermentation. A typical analysis of landfill gas produced during "steady-state" fermentation, is shown in Table 2.1. It should however be recognised that considerable variations in gas composition about the concentrations quoted are likely, depending on such factors as the age of the waste and the degree of microbial activity within it.

2.2 In addition to the major components referred to above, landfill gas contains a wide variety of components in trace amounts (Appendix I). Furthermore the gas is usually saturated with moisture which has been reported to contain constituents making it corrosive when in contact with metallic parts of any plant. Some of the minor constituents such as mercaptans are responsible for the characteristic odour associated with landfill gas. Although this odour may be a nuisance there is little evidence that it constitutes a significant health risk. Within a landfill, hydrogen sulphide is not normally present in concentrations exceeding 10 parts per million. However there is at least one example of a landfill in the UK where hydrogen sulphide concentrations exceeded 30% due to the presence of large quantities of material containing high concentrations of sulphate (e.g. gypsum waste, mixed with domestic and commercial waste). In this case the gas represented a considerable hazard. At low concentrations hydrogen sulphide is readily identified from its characteristic odour, however, potentially hazardous concentrations of hydrogen sulphide are not detectable by smell due to olfactory damage after only seconds of exposure. Arsine generation has also been reported when arsenic bearing cover material was used.

2.3 All the landfill gas constituents indicated both in Table 2.1 and Appendix I reflect the normal processes which occur in an operational or completed landfill. Landfill gas as measured within a plant may contain additional oxygen and nitrogen as a result of air being drawn into the recovery system installed within the landfill or via leaks or fractures in the gas pipelines or elsewhere in the plant.

Potential hazards of landfill gas

2.4 Landfill gas consists of a mixture of flammable, asphyxiating and noxious gases and may be a hazard to health. When air is permitted to mix with landfill gas, flammable mixtures may be produced which, if contained, can give rise to an explosion. The flammable range for methane is approximately 5% to 15% by volume in air,

and that for hydrogen approximately 4—74%. The presence of carbon dioxide affects the flammability limits (sometimes known as explosive or explosibility limits) of methane where the methane comes from a landfill gas containing 60% methane and 40% carbon dioxide.

2.5 The presence of carbon dioxide (density 1.5) increases the density of landfill gas over and above that of methane (density 0.5) and as a result it may be lighter or heavier than air. Gas liberated in a closed area, or even in the open in calm conditions, may not be uniformly mixed with air. As a result, stratification may enable flammable volumes of gas to collect and remain in buildings, structures, pipework or areas which were previously thought to be free from flammable gas concentrations.

2.6 In confined spaces such as culverts and manholes, gas accumulation can give rise to asphyxiation hazards.

2.7 Landfill gas may also pose particular problems due to corrosion or fouling of pipework.

Risk avoidance

2.8 Risks of fire, explosion, and asphyxiation may be avoided during operation, restoration and the post closure management period by:

(a) Siting plant handling flammable gas in the open air.

(b) Controlling sources of ignition in the vicinity of plant handling flammable gas. Electrical equipment should be sited in a safe place or be suitable for use in a hazardous area as recommended in BS 5345: Part I: 1976.

(c) Monitoring to ensure operations on a landfill are carried out only where the landfill gas concentration is within safe limits.

(d) Controlling gas extraction rates and monitoring abstraction pipes to ensure that any air is drawn in does not create potentially hazardous mixtures.

(e) Restricting entry into any shaft, culvert or other confined space where gas may accumulate. If access is considered to be essential various safety procedures should be considered and agreed before permission is granted. [See also BS 5573, 1978 which refers to safety issues in the construction of large diameter boreholes.]

(f) Ensuring that all personnel are aware of the potential risks on the surface of the site and avoid any practice which could lead to ignition. "No smoking" notices are obligatory and fires should not be lit.

TABLE 2.1 LANDFILL GAS COMPOSITION – OBSERVED RANGES IN THE UK
(expressed as % by volume)

Component	Typical value (mature refuse)	Observed maximum	Reasons for component being unusually abundant
Methane	63.8^2	77.1	Adsorption of carbon dioxide (e.g. by water; lime)
Carbon dioxide	33.6^2	89.3	Aerobic degradation of refuse
Oxygen	0.16^2	20.9^4	Air mixed with landfill gas
Nitrogen	2.4^2	80.3	Air mixed with landfill gas, or very slow degradation if oxygen depleted
Hydrogen	< 0.05	21.1	Young refuse. Methane concentration usually low
Carbon monoxide	< 0.001	$–^5$	Oxygen starved burning in refuse
Saturated hydrocarbons	0.005^3	0.074	Young refuse or high concentrations of petro-chemicals present
Unsaturated hydrocarbons	0.009^3	0.048	Young refuse or petrochemicals/solvents present
Halogenated compounds	0.00002^3	0.032	Young refuse or solvents present
Hydrogen sulphide	0.00002^3	0.0014 35^6	Young refuse High sulphate waste present
Organosulphur compounds	$< 0.00001^3$	0.028	Young refuse
Alcohols	$< 0.00001^3$	0.127	Young or semi-aerobic refuse (fermentation)
Others (not included above)	0.00005^3	0.023	Solvents or other volatile wastes deposited

Notes:

1   Landfill gas usually emerges saturated with water vapour, representing 0.001% to 0.004% depending on its temperature.

2   Based on long term data from Stewartby landfill, supplied by London Brick Landfill Limited.

3   Based on five year old refuse.

4   Entirely derived from atmospheric oxygen.

5   Concentrations of several percent carbon monoxide have been reported at landfills on fire but have not been confirmed.

6   Refuse mixed with plasterboard.

3. INITIAL SITE ASSESSMENT

3.1 Every landfill site should be evaluated for its potential impact on the surrounding environment from landfill gas migration or release. The assessment should cover possible odour problems from gas migration and release as well as the potential hazard to the safety of persons and buildings and the risk of damage to vegetation (see Chapter III, paragraph 3.75). At some of the larger sites the recovery of gas may be economically viable either independently of, or in conjunction with, a gas migration or odour control scheme. The flow diagram in Figure 3.1 indicates the various questions and actions that need to be asked or taken if an initial investigation indicates that landfill gas could give rise to problems at the site; the diagram is also relevant to the investigation of an incident in which landfill gas is implicated. As the flow chart indicates, unless the initial site assessment can guarantee that no hazard exists or is likely to exist after completion of the site, it will be necessary to introduce and operate a monitoring programme for as long as landfill gas could be produced (this aspect is discussed further below — see section 8). Various courses of action may be required depending on the severity of the problem identified, including the installation of a permanent gas control scheme. If such a scheme is required it will be necessary to decide whether it can take the form of passive venting (perhaps applicable to shallow sites up to about 5m depth) or whether a pumped gas control scheme is needed with wells or trenches within the waste (in deep sites).

3.2 Figure 3.2 indicates the stages involved in evaluating the potential of a site for the commercial utilization of landfill gas. It should be stressed that the decision to proceed, should the results of such an evaluation look favourable, will be based largely on economic factors related to the local circumstances of the proposed scheme. Many of the factors which make up the evaluation are inter-related and it is therefore not possible to give a definitive YES or NO answer to every statement or question. For example a small site may provide economically viable quantities of gas to a small local user to replace natural gas, whereas the same installation elsewhere would not be viable.

3.3 At most sites where a pumped gas control scheme is required or where commercial gas recovery is envisaged it will be necessary to carry out a field test programme. This involves the installation of a limited number of gas abstraction wells (or trenches) and periphery monitoring boreholes, to provide data on the efficiency of wells from which the size and scale of plant and equipment required for a permanent installation can be determined. Attention should be given at this field investigation stage to the various safety and legal requirements identified in later sections of this Annex. Drilling contractors should be notified of the potential hazards of landfill gas such as explosion, obnoxious gases and fire risk and of the need for protection of temporary on-site testing equipment. Having ascertained the parameters for the design of an effective scheme to control or utilize gas it will then be necessary to adhere to the various legal and other requirements for the full installation.

3.4 Where there is uncertainty or concern about the duration of gas production at any location within the landfill, laboratory testing of waste samples may provide more definitive information.

3.5 If an incident should occur involving any member of the public who believes that there is an escape of gas it is probable that the British Gas Corporation will be contacted first, at a local level. Thus whether or not the pipeline has been authorised under the Pipelines Act 1962 (see 4.11) and whether or not thereby the details of the pipeline have been notified to British Gas, it is essential for the safety of British Gas employees and the public in such a situation that the routing and locations of landfill gas pipelines installed off the landfill site are notified to the regional office of British Gas. The notification should include information on gas composition, whether it is heavier or lighter than air, any noxious or toxic properties, the system pressures, pipe materials, location and depth of the pipeline.

3.6 It may also be in the landfill site operator's own interest to advise the regional office of the British Gas Corporation of sites producing gas so that this can be taken into account if they are contacted by the public reporting an incident.

Annex 2: Figure 3.1.   DECISION FLOWCHART – GAS PROBLEMS

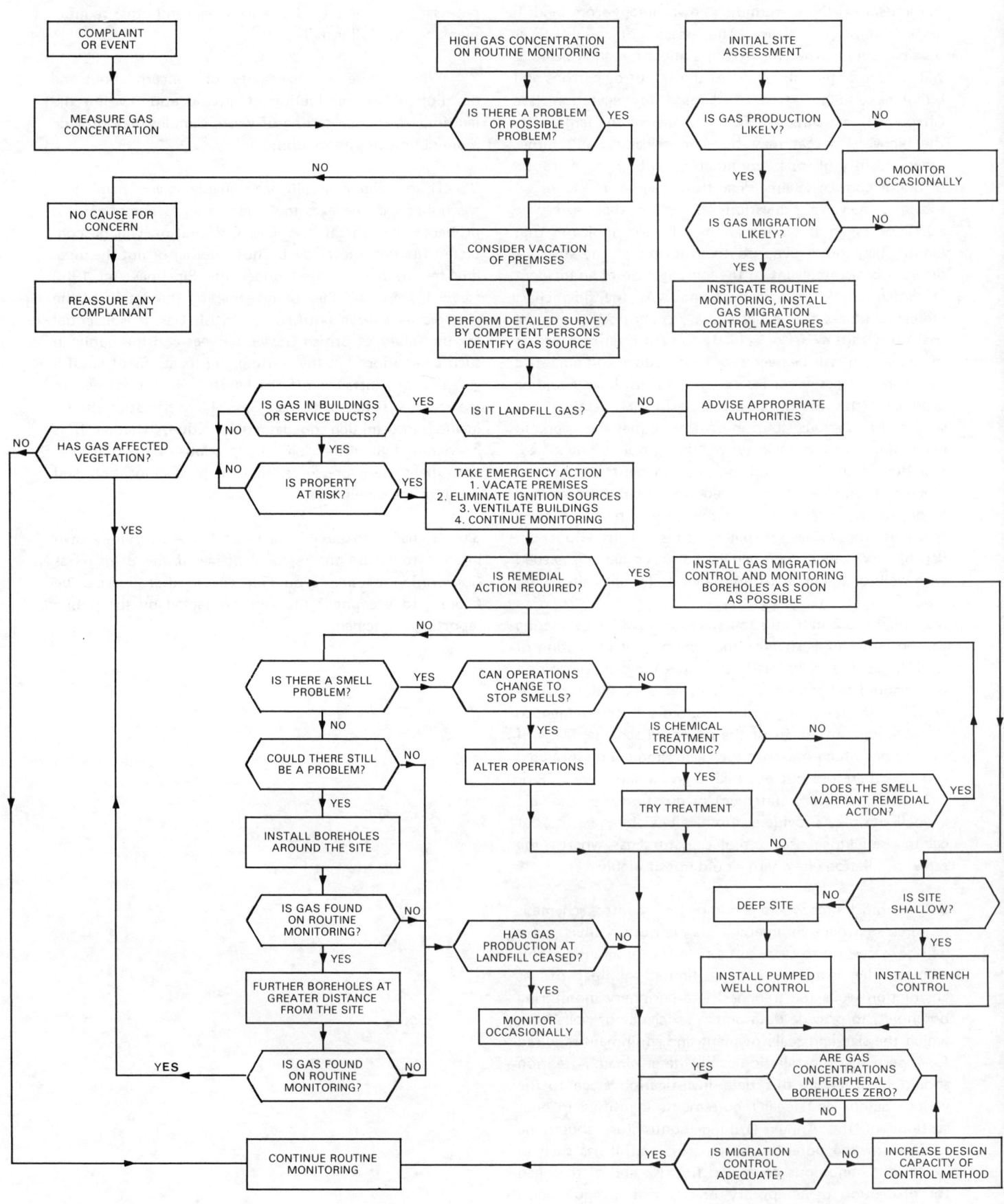

Annex 2:Figure 3.2. DECISION FLOW CHART — GAS UTILISATION

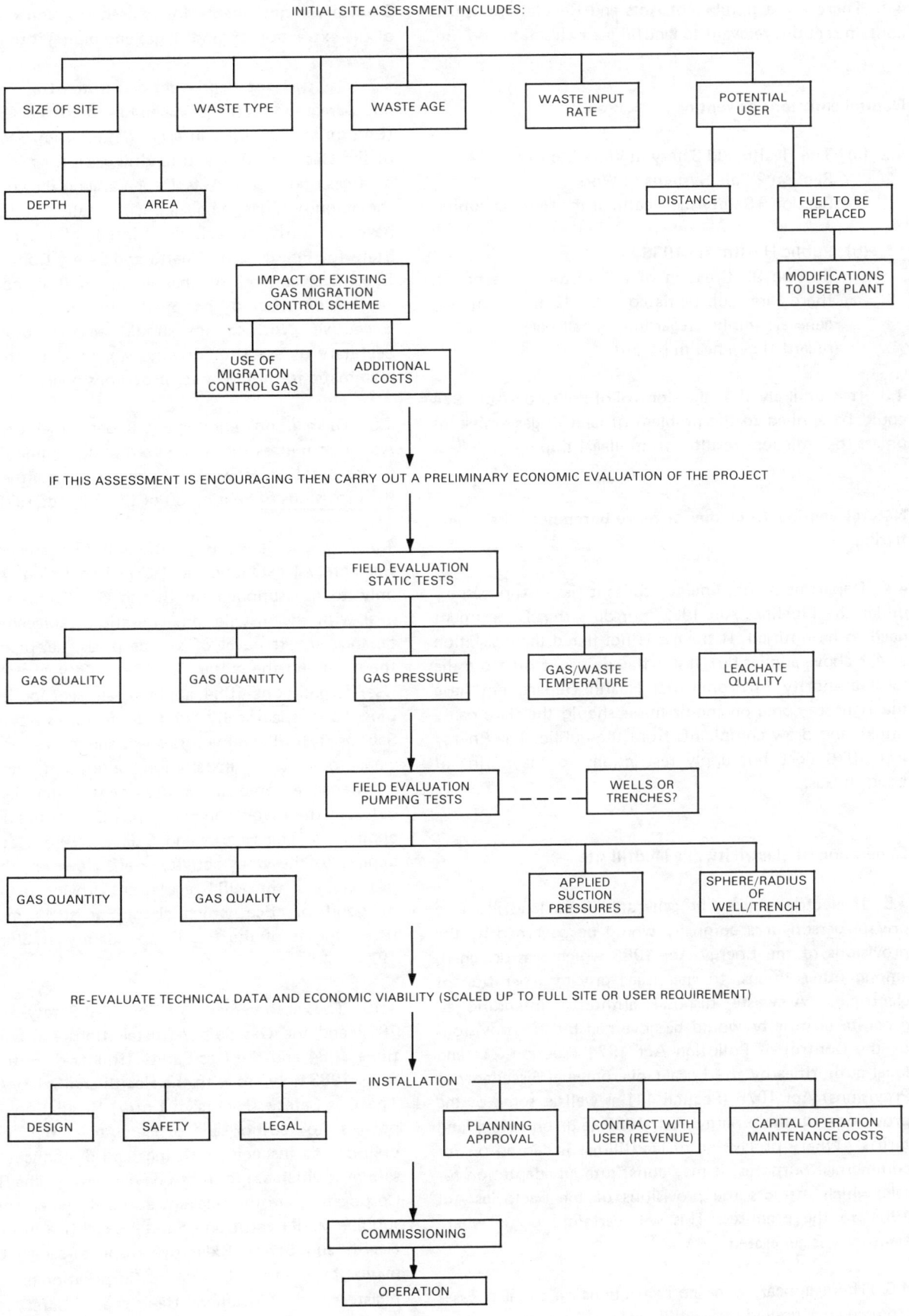

4. LAW RELATED TO LANDFILL GAS

4.1 There are a number of Acts and Regulations which contain sections relevant to landfill gas extraction and use.

Natural emissions or venting

4.2 (a) The Health and Safety at Work etc Act 1974
Section 2 Safe Systems of Work
Section 3 Safety and Health of the General Public

(b) Public Health Act 1936
Section 92 Creation of a Statutory Nuisance if, there are substantiated complaints from the general public regarding smell or, if gas is regarded as a health hazard.

4.3 It is unlikely that the Control of Pollution Act 1974 could be applied to the problem of landfill gas emissions unless the emissions resulted from illegal tipping.

Natural venting from one or more boreholes on site and flaring

4.4 Department of Energy consent is not necessary under the Pipelines Act 1962, nor does that Department need to be notified. If the gas is not flared the legislation at 4.2 above applies but, if it is flared, inspectors from the local authority Environmental Health Department have the right to come on the premises should the flare cause smoke and draw complaints from the public. The Energy Act 1976 does not apply restrictions to the flaring of Landfill Gas.

Generation of electricity at a landfill site

4.5 If electricity is to be generated from landfill gas, a private person or a company would be governed by the provisions of the Energy Act 1983 which was designed, among other things, to encourage private generation of electricity. A waste disposal authority intending to generate electricity would be governed by the provisions of the Control of Pollution Act 1974 (Section 21) and local authorities by the Local Government (Miscellaneous Provisions) Act 1976 (Section 11) as well as some of the provisions of the Energy Act 1983 (e.g. on safety and tariffs). Where processing of landfill gas is carried on for commercial purposes, it may constitute an adaptation for sale which attracts the provisions of the Factories Act 1961 to the premises. This will certainly occur where electricity is generated.

4.6 There appear to be no restrictions on exhaust gases from engines fuelled by landfill gas.

Statutory requirements for consent to, and management of the extraction of landfill gas and piping it to premises

4.7 Gas was not clearly defined in the Gas Act of 1972 but paragraph 17(b) of Schedule 3 to the Oil and Gas (Enterprise) Act 1982 inserts a definition in Section 48(1) of the Gas Act 1972 and landfill gas is now considered to be a gas under both Acts. Depending on the customer and the amount of gas to be supplied, a potential supplier may have to notify or seek the consent of the Secretary of State for Energy. The Health and Safety Commission and Executive act on his behalf in granting consent and application should be made to Health and Safety Executive. Notifications should be sent direct to the Secretary of State for Energy. Appendix IV shows in diagrammatic form how these provisions operate.

4.8 There is no requirement under either of the above Acts that meters should be used to determine the amount of gas supplied. However, if a meter is used it is subject to the provisions of Section 30 of the Gas Act 1972.

4.9 The Gas Safety Regulations 1972 as amended by the Gas Safety (Installation and Use) Regulations 1984 apply only to gas supplied by British Gas Corporation. They deal with the installation, alteration, replacement, maintenance and removal of service pipes, i.e. pipes between the main and the meter. The Gas Safety (Installation and Use) Regulations 1984 apply to all gas supplied through pipes but specifically exclude factories as defined in Section 175 of the Factories Act and mines. These regulations deal with installation, alteration, replacement, maintenance, removal and use of gas fittings, i.e. gas pipes between the meter and the appliance, fittings, meters and appliances. The Health and Safety at Work etc Act also applies to the work activity itself. However, the relevant provisions of the 1984 regulations provide for a standard of good practice which should assist in meeting the requirements of the Health and Safety at Work etc Act 1974.

4.10 The enforcement of the Gas Safety Regulations 1972 and the Gas Safety (Installations and Use) Regulations 1984 and the Gas Safety (Rights of Entry) Regulations 1983 now lies with the Health and Safety Executive. The Gas Safety (Rights of Entry) Regulations 1983 give powers to authorised officers to enter consumers' premises to inspect gas fittings and disconnect and make safe any which are found to be dangerous. The British Gas Corporation are the relevant authority when they are the supplier and Health and Safety Executive in other cases. Health and Safety Executive Inspectors may be accompanied by either a British Gas Corporation employee or a registered gas installer. Health and Safety Executive Inspectors will also deal with complaints from consumers.

Pipelines Act 1962

4.11 With certain exceptions, the Pipelines Act 1962 applies to all commercial pipelines to be constructed in England, Wales and Scotland. The construction of a cross-country pipeline (one more than 10 miles long) has to be authorised by the Secretary of State for Energy and the authorisation invariably carries with it deemed planning permission (Section 1 of the Act). Authorisation is not required for a local pipeline (one 10 miles long or less) but planning permission is, and notification has to be given to the Secretary of State not less than 16 weeks before the start of construction (Section 2 of the Act). Landfill gas pipelines will generally be less than 10 miles in length and hence be local pipelines under the Act.

4.12 The Act is administered by the Pipelines Inspectorate of the Department of Energy who also act as agents of the Health and Safety Executive on matters connected with pipeline safety. The statutory procedure for providing notification of intention to construct a local pipeline is set down in Section 2 of the Act but it is recommended that informal consultation takes place with the Pipeline Inspectorate at an early stage in the project. The Pipelines Inspectorate require the applicant to complete a technical data sheet a copy of which is attached at Appendix V, to provide essential information for certification of the pipeline.

4.13 The Health and Safety Executive may, at any time, impose safety requirements in respect of construction, testing, operation, maintenance or repair of a pipeline. These powers are exercised by means of a safety notice issued by the Pipelines Inspectorate on behalf of the Health and Safety Executive. Such a notice will require the pipeline to be designed, built, operated and maintained to an agreed standard with the broad aim of ensuring that the possibility of failure is negligible.

The Gas Quality Regulations 1983

4.14 These regulations prescribe standards of pressure, purity and smell to be complied with by any persons supplying gas through pipes. They also make provisions for furnishing information by persons, other than the British Gas Corporation, supplying gas through pipes and apply the Gas (Testing) Regulations 1949, with modifications to such persons. Other provisions deal with penalties, exceptions to liability and exemptions, and the revocation of Part 2 of the Gas Quality Regulations 1972. Landfill operators should notify the Gas and Oil Measurement Branch of the Department of Energy of their intention to supply gas. They may wish to apply for exemption from the requirements on pressure, smell or purity in the Gas Quality Regulations 1983 in respect of certain supplies for industrial purposes. A requirement of the regulations is that the gas should possess a distinctive smell. Gas as produced from the landfill normally has a distinctive smell, but if this is not sufficiently strong or the landfill gas has undergone some form of purification, it may be necessary to add a stenching agent.

Gas (Metrication) Regulations 1980

4.15 These Regulations amend the provisions of the Gas Act 1972 and Regulations made under it by substituting "for the definition of calorific value" a definition expressed in terms of metric units, for example the definition of "therm" expressed in terms of British Thermal Units is substituted by a definition expressed in terms of megajoules. The provisions of the Gas Safety Regulations 1972, the Gas Quality Regulations 1972 and the Gas (Meters) Regulations 1974 are amended by substituting for quantities expressed in imperial units, quantities expressed in metric units.

The Notification of Installations Handling Hazardous Substances Regulations 1982

4.16 These will not normally apply because the quantity of gas within the pipelines will normally be less than 15 tonnes and therefore insufficient for notification purposes. In addition pipelines authorised under section 1 of the Pipelines Act as notified under section 2 of the same Act are exempt from separate notification under the Notification of Hazardous Installations Regulations. If storage is proposed then the installation could be notifiable.

The Health and Safety (Flammable Gases Oxygen and Calcium Carbide) Regulations and Approved Code of Practice.

4.17 It is proposed that these regulations will apply to all flammable gases and would include the extraction and use of landfill gas and associated gas fittings. Their Approved Codes of Practice will have to be followed.

5. PLANNING PERMISSION

5.1 Currently it is understood that planning permission may not be required during the initial field testing phase of a gas control or utilization project. Planning authorities may require planning permission to be obtained for the testing of landfill gas for commercial viability where, for example, any flare-stack or large mobile plant may intrude on the public. Planning permission will be required for any permanent installation including a pumping station and flaring compound (if incorporated into the design). The permission may include the design and location of the flare-stack. Planning permission will be required for any local pipeline [where gas is to be delivered to a third party] and an application should be made to the Local Planning Authority at the same time as notification is given to the Pipelines Inspectorate of the Department of Energy.

5.2 Planning permission will also be required for siting a gas engine or turbine for the generation of electricity from landfill gas, regardless of whether the electricity is used on or off the site.

5.3 At new landfill sites it is likely that proposals for gas control will be required before planning permission for landfilling is given. If significant gas generation is foreseen within the life-time of any landfill site operation, conditions may also be applied within the terms of the disposal licence issued by the disposal authority. Activities such as sinking boreholes or excavating trenches at an operational site may be subject to the conditions specified by a disposal licence, which should be checked and advice sought from the disposal authority.

5.4 Notwithstanding the foregoing paragraphs in this section any proposed activity involving the exploitation of landfill gas should be discussed with the local planning authority. A procedure exists under Section 53 of the Town and Country Planning Act 1971 whereby an intending developer can obtain from the local planning authority a decision as to whether his proposal constitutes development for which planning permission is required.

6. PLANT AND EQUIPMENT

6.1 Landfill gas standards or recommendations go beyond those that would normally be required for similar plant in other circumstances, reflecting the particular problems posed by landfill sites, as follows:

(a) A landfill site will produce gas which may migrate beyond the site boundary, or vent naturally through the surface of the site.

(b) The gas may be particularly corrosive or noxious, as well as having flammable and asphyxiant properties.

(c) The amount of gas produced at any time may be greater than in other anaerobic digestion systems for which plant and safety standards already exist.

(d) Weather conditions can exacerbate the problems associated with the gas.

(e) The site may be operational in terms of continued waste disposal, during the period of gas production.

(f) The gas cannot be turned off.

6.2 Risks can be minimised by the use of purpose built equipment and apparatus which has been designed to meet the required British Standards or other acceptable professional recommendations of existing gas handling industries. The use on landfill sites of makeshift equipment including, in particular, gas flares is unacceptable.

6.3 The gas collection facilities and associated plant and pipework should be designed, constructed and installed so as to avoid the risk of either gas leakage, or excessive air ingress into the landfill or plant. The plant, pipework and fittings should be suitable for the pressures and temperatures which might arise. Pipework should be tested to 1.5 times the maximum allowable operating pressure or under full vacuum as appropriate. Relief devices should be fitted on all positive displacement machines. All valves, vents and any safety relief devices should be arranged to discharge at a safe position through a flame arrestor to the open air.

6.4 A considerable amount of legislation, standards, codes of practice and other safety documents exist (see Appendices II and III) with sections which are applicable to the drilling, installation and operation of gas collection facilities and plant on landfill sites. However, this information has not in general been developed for landfill operations. Areas where additional care is needed reflect the unique aspects of landfill site operation and relate to deposit of wastes, vehicle access, drilling, the corrosive nature of the gas, aspects of environmental effects of flaring and prevention of vandalism.

6.5 The remaining paragraphs in this section describe various aspects of plant design and operation which must be considered in connection with gas extraction and preparation for use on site or for sale to an external user and must be considered from the following aspects.

Overall site operations

6.6 The site should be operated in such a way as to minimise the possibility of accidental damage to wells or pipes by vehicles working on the site. Where available, the history of the site and details of material deposited should be considered in order to anticipate unusual gas composition, hazards etc.

Drilling of wells

6.7 Drilling of monitoring boreholes or extraction wells will locally accelerate any escape of gas. Consideration should be given to the possibility of release of noxious, toxic and/or flammable gas during drilling. For these reasons it is recommended that, during drilling, all wells should be cased to minimise gas release. During drilling regular checks for gas quality should be made (see section 8). Factors leading to increased risk at the landfill surface include calm air, wind shelters such as hollows and spaces under vehicles. Personnel should not descend into pits or boreholes unless suitably equipped (see 2.8(e)). All drilling contractors must be made aware of potential hazards and be suitably advised and supervised. At no time must fewer than 2 people be present.

6.8 Where test wells will not be permanent structures they may be constructed of either plastic or steel. Permanent extraction wells should, however, preferably be constructed using a plastic such as uPVC, HDPE or polypropylene, (see Table 4.54), since landfill gas will almost certainly contain trace constituents corrosive to steel unless it is suitably protected or is corrosion resistant.

6.9 Typically, extraction wells (see figure 6.9) which have a diameter of between 0.3 to 1.0 metres are drilled almost to the base of the refuse or until the landfill water table is reached. The well pipe, usually some 0.1-0.15 m in diameter, perforated except for the top few metres, is inserted into the drilled hole. The annulus is backfilled with permeable material such as gravel, rock or broken bricks, and is capped by a layer of impermeable material. The head of each well may be fitted with a flow/pressure control valve and suitable monitoring points for flowrate, pressure and temperature measurements if required. These measures may be particularly important in a gas migration control system. In some installations, where individual

Annex 2: Figure 6.9. LANDFILL GAS EXTRACTION WELL

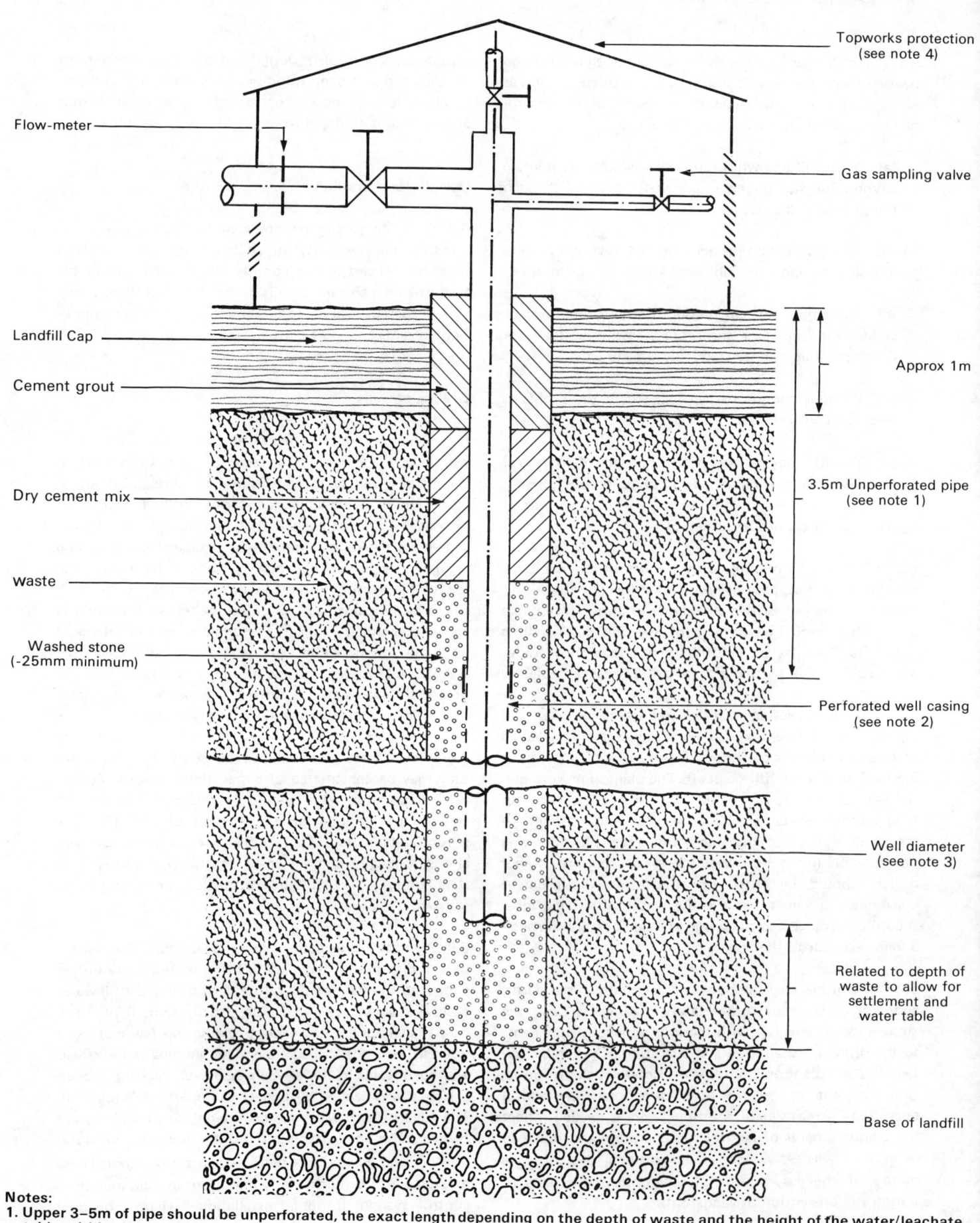

Topworks protection (see note 4)

Flow-meter

Gas sampling valve

Landfill Cap

Cement grout

Approx 1m

Dry cement mix

3.5m Unperforated pipe (see note 1)

waste

Washed stone (-25mm minimum)

Perforated well casing (see note 2)

Well diameter (see note 3)

Related to depth of waste to allow for settlement and water table

Base of landfill

Notes:
1. Upper 3–5m of pipe should be unperforated, the exact length depending on the depth of waste and the height of the water/leachate table within the waste.
2. The remaining pipe should be perforated, (approx. 20% perforated area) preferably made of uPVC, Hdpe or polypropylene and of at least 110mm diameter.
3. Overall well diameter depends on the amount of landfill gas to be extracted from the well. This should be determined by static tests on trial wells and in most circumstances,should preferably be a minimum of 300mm diameter.
4. Topworks may be suitably protected above ground, by a robust ventilated enclosure or below ground, in a manhole which does not compromise the integrity of the landfill cap.

wells are linked to a central manifold, these measurements and controls may be carried out at this location.

Pipework and flame arrestors

6.10 Gas production wells should be joined together by plastic pipe, usually of at least 100 mm diameter, which is in turn connected to a suction pump. These pipes, which can be laid on the surface of the landfill or buried should be sloped or so designed to facilitate the drainage of condensate back to the well or to other points for its removal. It is important that no low spots are present in this pipework.

6.11 When excavating trenches for inserting gas abstraction pipes similar precautions should be taken as for drilling wells. The potential hazards are greater and open trenches should therefore be kept as small as possible and backfilled on completion of installation work. Because of the possibility of refuse instability shoring may be required at any depth and will be obligatory at depths greater than 2 m. Laying of pipes should be carried out from outside the trench whenever possible. Men should not normally enter such trenches, but if entry is unavoidable, they must be protected from gas. Whenever practical, the use of horizontal gas collection pipes should be confined to their being laid on previously deposited waste. The use of metal pipes in waste as either vertical or horizontal gas collection lines is not recommended as they will corrode. The plasticiser in some plastic pipes can react with light, causing enbrittlement, and some grades of PVC are susceptible to softening (and hence collapse) at temperatures as low as 45° celsius which frequently can be reached in landfilled waste.

6.12 A flame arrestor, or water seal trap is essential between a pump and a flare stack. Experience indicates that regular maintenance of these flame arrestors is essential due to clogging by sand, dust and water. Some refrigerated cooling of gas may be desirable or essential depending on end-use.

Plant, gas boosters and pumps

6.13 The main items comprising a gas abstraction plant are usually a gas moisture removal trap, an extractor pump, associated pipework, valves, control and monitoring equipment and power supplies and possibly a gas cooler. In addition a flare stack may be required.

6.14 An important plant item is some form of moisture knockout trap or chiller unit. The efficiency required of such a trap and its position within the plant will depend on the pump/blower/compressor being used. Such traps also may be required within the gas collection system in the landfill and should be designed to be safe from implosion under negative pressure and contain adequate drain-

off facilities which will avoid air intrusion during operation.

6.15 The distribution network of pipes and controls required at each plant will be unique although design of such systems should follow standard practice for gas handling plant. Some plants contain gas engines or turbines to convert landfill gas into electrical power. Such plants may need substantial electrical sub-stations, protected from gas ingress, to provide interconnection with the national grid.

6.16 The following sub sections indicate the types of blower or compressor which are suitable for various landfill gas extraction applications. It is noted that there are some types of gas movers, which by virtue of their sealing arrangements are not suitable for landfill gas.

(a) *Centrifugal blower*

This machine consists of an impeller with a series of radial vanes which rotate within a specially shaped casing. Fluid is drawn into the hub (or eye) and is thrown outwards by centrifugal force. Static pressure increases as the fluid moves outwards. Pressure is further increased in the diffuser section of the casing, where fluid velocity is converted to 'pressure energy'. This type of machine is suitable for supplying large volumes of gas at reltively low pressures.

(b) *Rootes type blower*

Specially lobe shaped impellers rotate within a casing and entrain fluid as they do so. The fluid is not compressed as it moves through the machine itself as the supply pressure depends entirely on the 'restriction' placed downstream from the blower.

(c) *Liquid ring pump*

This is a rotary displacement type of machine which is sealed by a 'liquid ring'. The liquid used is usually water and the ring is formed by centrifugal forces which arise when an eccentrically mounted impeller rotates within an irrigated enclosure.

The service liquid has three functions:—

(i) Sealing between impeller and casing. Mechanical clearances need not, therefore, be close.

(ii) Heat produced during the compression is removed and can be 'dissipated' elsewhere.

(iii) Power from the motor is absorbed and transmitted to the gas being compressed.

Liquid ring pumps are especially suited to pumping dirty or moisture laden gases and they can operate with minimal maintenance.

(d) Sliding vane rotary compressors

This type of machine consists of an eccentrically mounted rotor with slots around the edge which support sliding vanes. As the rotor revolves, it causes the vanes to slide in and out of the slots, and the tips of the vanes make contact with the casing by centrifugal force to provide an effective seal. The volume between rotor, casing and the vanes changes during rotation and leads to compression of the gas stream. A non-return valve downstream of the compressor provides protection against flow reversal which could occur when the machine is unloaded.

(e) Screw compressor

Two helical rotors revolve in mesh and entrain and compress gas. There is no contact between the rotors or between rotors and casing, hence lubrication is not required and the delivery is not contaminated. Both rotors are driven from a single motor through a timing gearbox. This type of machine is capable of providing high pressures.

(f) Reciprocating pumps

These involve the reciprocating motion of a piston within a cylinder. The fluid is drawn into the cylinder through an inlet valve and after compression, leaves through a delivery valve. Because the compression process is not continuous, the delivery 'pulsates'. A more constant discharge from this type of machine can be obtained by interposing a receiver between the pump and the pipeline. Reciprocating machines are large and heavy, but they are capable of producing very high pressures.

(g) Ejectors

A simple pump type with no moving parts. Suction head is provided by fluid passing rapidly through a venturi. The pump has three basic parts:— (1) nozzle, (2) suction chamber and (3) mixing diffuser (venturi). The 'working fluid' is expended through the nozzle and mixes completely with the landfill gas being pumped. The working fluid is usually compressed air but could be steam. They both have the disadvantage that the gas is diluted by the working fluid which can impair flaring. Recirculating liquid ejectors can overcome this problem.

6.17 Appendix VI indicates the maximum delivery pressures of the pumps described together with their relative maintenance costs and the types of applications in which they might be used.

Flares

6.18 Flare stacks are not always essential and simple venting may be sufficient in many instances. However, many site specific factors, and environmental requirements such as odour control, may make their use essential. They should be installed at a high level to limit the environmental nuisance by emissions from the stack top. The height required will depend on local circumstances, the main concern being to achieve adequate dilution of the gas if loss of ignition occurs and to dilute the products of gas combustion. Radiant heat from the burning gas can be substantial. Where a flare stack is required and in particular where it forms part of a gas migration control system, some form of monitoring of the gas stream or flare temperature sensor will be required. This monitoring device must be connected to an alarm and call-out systems so that repairs can be effected or stand-by equipment can be brought into use immediately.

Siting and security

6.19 The gas pumping equipment should be sited in a safe well ventilated position. An open air location is recommended as far from residential properties as is possible. If weather protection is required, a Dutch barn type of structure made of non-combustible material should be used. If further weather protection is required at least two opposite sides of the structure should be left open for ventilation and explosion relief. Where gas compressors have to be sited in poorly ventilated areas, such as within buildings, forced ventilation should be provided, and the atmosphere should be monitored for methane. The detectors should be interlocked to isolate the power supply to the compressors in the event of a gas leak. Manual re-start with automatic purging is required. The building should be provided with explosion relief.

6.20 The compound will commonly be located near to the landfill, particularly where the user is a distance away. With high volumes of gas to be transmitted, pressurisation is desirable to keep down the size, and hence the costs, of the pipeline. Location of the pumping compound at or near the landfill may not be feasible and location at the user end may be adopted. This course may be preferable in other cases for example where a site is surrounded by property and plant noise cannot be sufficiently reduced by sound insulation.

6.21 All sources of ignition, for example, naked flames, hot surfaces, smoking materials etc must be eliminated from areas in the vicinity of plant handling flammable gases. Personnel should be aware of the possibility of ignition of gas anywhere on the site. Wherever possible electrical equipment should be located in a safe place remote from potential sources of leakage of flammable gas. Electrical equipment which has to be located in a hazardous area should be suitably protected. A code of practice on electrical apparatus for use in potentially explosive atmospheres is included in British Standard 5345 1976 (Appendix II.19). Safety and warning notices etc should be posted where there is a possibility of a gas leak.

Colour coding and labelling

6.22 All pipes should be colour coded. The run of underground pipes should be indicated clearly by above ground markers, and by marker tapes laid in the ground, approximately 0.3 metres above the pipework. Plant and apparatus should be labelled and a plant schematic and valve identification chart should be displayed for use by the operators. Emergency procedures should be listed and prominently displayed.

7. INSTRUMENTS FOR TESTING AND CONTROL

Methane

7.1 Many portable methane detectors are available. Although they are not designed for use with landfill gas they usually give reasonably accurate data on methane concentrations, but in some cases they may not. Therefore, it is recommended that some gas samples be collected and analysed by gas chromatography, not only to confirm the recorded methane composition, but also to provide regular calibrations of such portable equipment. Similar arguments apply to the detection of other major constituent gases commonly found in landfill gas such as hydrogen, oxygen, nitrogen and carbon dioxide.

7.2 Most portable instruments use a pellistor catalytic detector (for the sensitive LEL ranges) and/or a thermal conductivity detector for the 0-100% gas by volume ranges. Portable infra-red detector instruments are available for both methane and carbon dioxide but are more expensive. An instrument for measuring up to 1% methane in air is available which uses flame ionisation detection. Although the flame will be extinguished at methane concentrations greater than 1%, care should be taken when using the instrument if such gas atmospheres are expected. In on-line plants instrument levels will be set, as appropriate, on a site-by-site basis. Back-up equipment should be available when servicing, or failure of detectors, occurs. Pellistor catalytic detectors are subject to poisoning.

Oxygen

7.3 Portable oxygen meters are suitable for spot check readings but some, if used continuously, will lose their accuracy due to moisture/corrosion/poisoning problems. In commercial applications oxygen detectors should be set to alarm at 4% and activate plant shut-down at a maximum of 8%. Additional oxygen monitoring equipment should be available when recalibration or servicing of instruments is necessary.

Gas analysers

7.4 In full-scale plants the use of on-line gas analysers is recommended for methane and oxygen monitoring. Occasional readings for both carbon dioxide and nitrogen would be desirable for general site control. In practice these would often be obtained by gas chromatography. Equipment using laser technology is being developed.

Calorimeters

7.5 Calorimeters are used to measure calorific value. Because corrosion problems can occur, suppliers should be consulted about the gas composition to be evaluated.

Atmospheric monitoring within plant buildings

7.6 Serious and fatal accidents have occurred in buildings and other confined spaces when concentrations of flammable and/or explosive gases have risen above the lower explosive limit. To prevent this happening it is essential that the concentration of methane in air is not allowed to rise above one quarter of its lower explosive limit (equivalent to 1.25% methane in air). However the rate of emission of gas from a landfill and its associated collection system will be influenced by uncontrollable factors such as climatic conditions, with the possibility of an unpredicted rise of the methane concentration in buildings. This potential for unexpected variation therefore demands that the alarm and power isolation trigger concentrations for methane in the atmosphere within plant buildings be set relatively low. Suitable methane detectors should be sited near potential sources of gas leakage or collection and set to give an alarm at 0.25% methane and to trigger isolation of power to the installation at 0.5% methane. Only manual restarting of the plant should be permitted and only then when the source of gas escape has been located, remedial action taken, residual concentrations eliminated and it has been certified safe to do so. This can be achieved by adequate forced ventilation, the precise rate necessary being calculated with reference to the anticipated maximum rate of evolution of methane. The adequacy of the ventilation should be assessed by testing the atmosphere with a correctly calibrated explosimeter. Further information may be found in Guidance Note GS5 from the Health and Safety Executive — "Entry into confined spaces".

Temperatures and pressure

7.7 Monitoring is advisable for flow metering corrections. This may be essential for some types of positive displacement pumps and advisable on dry gas distribution systems. Both temperature and pressure monitoring within the gas field, relayed back to the plant, may be desirable to provide warning of dramatic changes within the landfill.

Flow metering

7.8 Metering of gas flow is essential when conducting site evaluations, for migration control or commercial potential, and obligatory when selling to a third party, and beneficial when extracting for environmental control.

7.9 Because landfill gas is generally handled wet and at

low pressure, consistent accurate metering is not achievable at low cost. Meters commercially available fall into four classifications:

(i) Differential pressure, e.g. orifice plate, nozzle and pitot tube;

(ii) Variable area;

(iii) Mechanical, e.g. positive displacement meters including Rootes type, and turbines;

(iv) Electromagnetic and ultrasonic.

7.10 Examples of each group and their applicability to landfill gas are reviewed in Appendix VII.

8. SITE MONITORING

8.1 At all sites where permeable strata surround the waste, a series of appropriately spaced monitoring boreholes should be provided to permit the detection of landfill gas migration from the waste. At new sites such boreholes should be installed prior to waste deposition, to allow background concentrations to be measured, and to warn of the onset of gas migration. Measuring gas concentrations in such systems should be undertaken with care as the volumes of gas present may be small. Air intrusion through the borehole top should be avoided by limiting the rates of gas withdrawal to the measuring instrument. Gas may be migrating at discrete horizons in the strata and localised depth sampling techniques may be required. The effects of changes in atmospheric pressure or water table levels may be important; such changes may cause fluctuations in the rate of gas migration from the site which may not be immediately detected at the monitoring point. Both oxygen and methane concentrations should be measured using portable apparatus with occasional more detailed analyses to confirm the data.

8.2 The water authority, and the disposal authority may require at least one well to be installed in the waste for monitoring leachate quality and levels. These wells might subsequently be used for leachate abstraction. In many cases such monitoring points consist of large concrete rings with perforations over at least the lower portion of the chamber. Because of the problems of gas escaping from these chambers it is recommended that the cover and upper sections be sealed to prevent localised high concentrations of gas escaping from the well. If these chambers are also used to monitor gas quality and quantities they should be fitted with a suitable, valved, monitoring point. Recent experience of the use of concrete rings for monitoring wells has shown that they may be attacked by leachate. For gas monitoring purposes it is preferable to install a proper well within the waste after final restoration levels have been reached which may also be used for gas abstraction. Suitable safety precautions should be taken when sinking wells into waste. Such wells should contain a central perforated tube the diameter of which should not exceed 0.15 m, thus not permitting personnel entry, and should be sealed at the surface and provided with a valve. Such wells would enable leachate monitoring to be carried out. Well design should take into account the possibility of settlement and lateral movement of the waste.

8.3 When monitoring, care should be taken to use suitable instruments such as 'non-flame' detectors. Staff taking readings must avoid direct contact with gas plumes at these monitoring points and observe the various safety precautions described above.

8.4 Where it is suspected that gas migration may be causing problems in buildings, service ducts, etc or posing a risk of nuisance or damage to vegetation the monitoring procedures described in the following paragraphs should be adopted (these procedures are summarised in figure 1; see paragraph 3.1).

8.5 The concentrations of gas within enclosed spaces should be ascertained initially by using portable explosion-proof methane detectors, before more sensitive detectors, perhaps incorporating flame ionisation techniques, are used. Sources of gas intrusion into, or build-up within, buildings include points where services enter or leave the building, cavity walls and normally enclosed spaces such as cupboards. Various external monitoring points may include drains, ducts, electrical and telephone service points. Gas concentrations at such locations can fluctuate from time to time due to variations in climatic conditions, gas generation rates and other factors. Therefore, complete reliance should not be placed on a single set of recorded data from one monitoring exercise. This also applies to the monitoring of gas in boreholes or wells within or near the site as described below.

8.6 One further complication is identification of the source of gas. In some circumstances methane gas may be from fossil fuel sources such as coal mines or natural gas. Gas samples should be obtained and where possible the source 'fingerprinted' by identification of constituents. Examples of typical gas compositions where the gases are derived from sources other than a landfill are shown in Appendix VIII.

8.7 Monitoring with respect to known or anticipated vegetation damage caused by landfill gas can sometimes be carried out by measuring gas quality in cracks in the soil, but more commonly the installation of probes extending into the plant root zone and the extraction of gas from such probes is necessary.

8.8 If gas migration is suspected but has not been positively confirmed sufficient boreholes must be installed, extending to the full depth of the waste and in the strata surrounding the site, to identify the extent of any unacceptable gas concentrations. This might require boreholes at close spacings and/or at increasing distances from the landfill boundary especially if the strata are particularly non-uniform in composition and perhaps, are fissured.

8.9 Useful data can be obtained from wells installed within sites, which may help in the interpretation of the behaviour of a site with respect to gas generation. Such data include: gas composition and rates of evolution; gas pressure and temperature; the temperatures and composition of leachate if present in the drilled well. Chemical and microbiological analysis of the waste material extracted during drilling may be useful in determining the status of gas production. Monitoring during field testing will include gas composition, temperature and rates of abstraction.

8.10 Gas migration control systems should include the monitoring of peripheral boreholes for landfill gas concentration. Odour identification can also be carried out if necessary and the atmosphere may be monitored to determine ambient air concentrations after the gas has been collected and flared.

8.11 Monitoring of installed gas control or utilization schemes will form part of the operation of the plant and equipment. However, it should be realised that each well or trench will not only have individual characteristics, but that rates of gas abstraction will vary with time as the decomposition of landfilled waste proceeds. Thus, some resetting of well head conditions (flow/pressure) using the installed valve controls will be inevitable.

8.12 If property had been affected by landfill gas, prior to any control measures becoming operational, continued monitoring on a regular basis is essential to confirm that the remedial measures taken continue to be effective.

9. PLANT DESIGN COMMISSIONING AND MAINTENANCE

9.1 The design of a plant will depend on local circumstances, the various safety regulations and should adhere to manufacturers equipment specifications (Appendices II and III). Similarly start up procedures and maintenance requirements should be specified together with staffing levels. The procedures to be adopted in the event of equipment failure or emergency should also be specified.

10. RATEABLE VALUE OF INSTALLED GAS PLANT

10.1 Any permanent installation may be liable for rating by the local authority, but is subject to negotiation with the Local Rating Authority.

Compounds detected in landfill gas from six UK sites

Compound	Chemical formula	Molecular weight	Concentration range observed (mg/m^3)
ALKANES			
Propane	C_3H_8	44	$<$ 0.1–1
Butanes	C_4H_{10}	58	$<$ 0.1–90
Pentanes	C_5H_{12}	72	1.8–105
Hexanes	C_6H_{14}	86	1.3–628
Heptanes	C_7H_{16}	100	4–1054
Octanes	C_8H_{18}	114	8.5–675
Nonanes	C_9H_{20}	128	31–226
Decanes	$C_{10}H_{22}$	142	81–335
Undecanes	$C_{11}H_{24}$	156	12–164
ALKENES			
Butadiene	C_4H_6	54	$<$ 0.1–20
Butenes	C_4H_8	56	$<$ 0.1–90
Pentadienes	C_5H_8	68	$<$ 0.1–0.4
Pentenes	C_5H_{10}	70	$<$ 0.5–2
Hexenes	C_6H_{12}	84	$<$ 0.5–136
Heptadienes	C_7H_{12}	96	$<$ 0.1–1.9
Heptenes	C_7H_{14}	98	0.3–103
Octenes	C_8H_{16}	112	$<$ 1–144
Nonadienes	C_9H_{16}	124	$<$ 0.1–9
Nonenes	C_9H_{18}	126	5.2–75
Decenes	$C_{10}H_{20}$	140	13–188
Undecenes	$C_{11}H_{22}$	154	$<$ 2–54
CYCLOALKANES			
Cyclopentane	C_5H_{10}	70	$<$ 0.2–6.7
Cyclohexane	C_6H_{12}	84	$<$ 0.5–103
Methyl Cyclopentane	C_6H_{12}	84	$<$ 0.1–79
Dimethyl Cyclopentanes	C_7H_{14}	98	0.1–330
Ethyl Cyclopentane	C_7H_{14}	98	$<$ 0.1–$<$2
Methyl Cyclohexane	C_7H_{14}	98	1.5–290
Trimethyl Cyclopentanes	C_8H_{16}	112	$<$ 0.1–58
Dimethyl Cyclohexanes	C_8H_{16}	112	$<$ 2–54
Trimethyl Cyclohexanes	C_9H_{18}	126	$<$ 0.1–27
Propyl Cyclohexanes	C_9H_{18}	126	$<$ 0.5–8
Butyl Cyclohexanes	$C_{10}H_{20}$	140	$<$ 0.1–4
CYCLOALKENES			
Limonene	$C_{10}H_{16}$	136	2.1–240
Other terpenes	$C_{10}H_{16}$	136	14.3–311
? Menthene	$C_{10}H_{18}$	138	$<$ 0.1–29

Compound	Chemical formula	Molecular weight	Concentration range observed (mg/m^3)
AROMATIC HYDROCARBONS			
Benzene	C_6H_6	78	0.4—114
Toluene	C_7H_8	92	8—>460
Styrene	C_8H_8	104	<0.1—7
Xylenes	C_8H_{10}	106	34—470
Ethyl Benzene	C_8H_{10}	106	17—330
Methyl Styrene	C_9H_{10}	118	<0.1—15
Propyl Benzenes	C_9H_{12}	120	36—292
Butyl Benzenes	$C_{10}H_{14}$	134	5.8—138
Pentyl Benzenes	$C_{11}H_{16}$	148	0.4—17.5
HALOGENATED COMPOUNDS			
Chloromethane	CH_3Cl	50	<0.1—1
Chlorofluoromethane	CH_2ClF	68	<0.1—10
Dichloromethane	CH_2Cl_2	85	<0.1—190
Chlorodifluoromethane	$CHClF_2$	86	<0.1—16
Dichlorofluoromethane	$CHCl_2F$	103	<0.1—93
Chloroform	$CHCl_3$	119	<0.1—0.8
Dichlorodifluoromethane	CCl_2F_2	121	<0.1—48
Trichlorofluoromethane	CCl_3F	137	<0.1—20
Chloroethane	C_2H_5Cl	64	<0.1—46
1,1-Dichloroethane	$C_2H_4Cl_2$	99	<0.1—130
1,2-Dichloroethane	$C_2H_4Cl_2$	99	<0.1—8
Vinyl Chloride	C_2H_3Cl	62	<0.1—32
1,1,1-Trichloroethane	$C_2H_3Cl_3$	133	<0.1—177
1,2-Dichloroethylenes	$C_2H_2Cl_2$	97	<0.1—302
Trichloroethylene	C_2HCl_3	131	<0.1—170
Tetrachloroethylene	C_2Cl_4	166	<0.1—350
1,1-Dichlorotetrafluoroethane	$C_2Cl_2F_4$	171	<0.1—1
1,2-Dichlorotetrafluoroethane	$C_2Cl_2F_4$	171	<0.1—10
1,1,1-Trichlorotrifluoroethane	$C_2Cl_3F_3$	187	<0.1—70
Bromothene	C_2H_5Br	109	<0.1—<2
Chloropropanes	C_3H_7Cl	78	<0.1—<2
Dichlorobutenes	$C_4H_6Cl_2$	125	<0.1—<2
Chlorobenzene	C_6H_5Cl	112	<0.1—2.1
Dichlorobenzenes	$C_6H_4Cl_2$	147	< 2—16
ORGANSULPHUR COMPOUNDS			
Carbonyl Sulphide	COS	60	<0.1—1
Carbon Disulphide	CS_2	76	<0.1—2
Methanethiol	CH_4S	48	<0.1—87
Ethanethiol	C_2H_6S	62	<0.1—<2
Dimethyl Sulphide	C_2H_6S	62	<0.2—60
Dimethyl Disulphide	$C_2H_6S_2$	94	0.1—40
Diethyl Disulphide	$C_4H_{10}S_2$	122	<0.1—0.6
Butanethiols	$C_4H_{10}S$	90	<0.1—2.4
Pentanethiols	$C_5H_{12}S$	104	<0.1—?1.2

Compound	Chemical formula	Molecular weight	Concentration range observed (mg/m^3)
ALCOHOLS			
Methanol	CH_4O	32	$<0.1-210$
Ethanol	C_2H_6O	46	$<0.1->810$
Propan-1-ol	C_3H_8O	60	$<0.1-110$
Propan-2-ol	C_3H_8O	60	$<0.1->46$
Butan-1-ol	$C_4H_{10}O$	74	$<0.1->19$
Iso-Butan-1-ol	$C_4H_{10}O$	74	$<0.1->5.3$
Butan-2-ol	$C_4H_{10}O$	74	$<0.1-210$
ESTERS			
Ethyl Acetate	$C_4H_8O_2$	88	$<0.1-64$
Methyl Butanoate	$C_5H_{10}O_2$	102	$<0.1-15$
Ethyl Propionate	$C_5H_{10}O_2$	102	$<0.1-136$
Propyl Acetate	$C_5H_{10}O_2$	102	$<0.1-50$
Isopropyl Acetate	$C_5H_{10}O_2$	102	$<0.1-?6$
Methyl Pentanoate	$C_6H_{12}O_2$	116	$<0.1-22$
Ethyl Butanoate	$C_6H_{12}O_2$	116	$<0.1-350$
Propyl Propionate	$C_6H_{12}O_2$	116	$<0.1-200$
Butyl Acetate	$C_6H_{12}O_2$	116	$<0.1-60$
Ethyl Pentanoate	$C_7H_{14}O_2$	130	$<0.1-27$
Propyl Butanoate	$C_7H_{14}O_2$	130	$<0.1-100$
ETHERS			
Dimethyl Ether	C_2H_6O	46	$0.02-<2$
Methyl Ethyl Ether	C_3H_8O	60	$<0.1-<2$
Diethyl Ether	$C_4H_{10}O$	74	$<0.1-12$
Dipropyl Ethers	$C_6H_{14}O$	102	$<0.1-220$
OTHER OXYGENATED COMPOUNDS			
Acetone	C_3H_6O	58	$<0.1-?3.4$
1,3-Dioxolane	$C_3H_6O_2$	74	$<0.1-?5$
Butan-2-one	C_4H_8O	72	$0.4-38$
Tetrahydrofuran	C_4H_8O	72	$<0.1-<2$
Pentan-2-one	$C_5H_{10}O$	86	$<0.1-4.2$
Methyl Furans	C_5H_6O	82	$<0.1-0.8$
Dimethyl Furans	C_6H_8O	96	$<0.1-12$
? Camphor/Fenchone	$C_{10}H_{16}O$	152	$<0.1-?13$
Carboxylic Acids	$C_nH_{2n}O_2$	–	$<0.1-<2$

Documents, standards and codes of practice relevant, at least in part, to landfill gas

1. A guide to the gas safety regulations. Report IM/4.1975. British Gas.

2. A code of practice on safety in and around anaerobic digesters. 1982. BABA Ltd.

3. Guidance on the design and commissioning of installations in potentially explosive atmospheres associated with sewage handling and effluent treatment plants. (Draft — in preparation). North West Water Authority.

4. Opening of gas plant and working in confined spaces. IGE/SR/S.1978. The Institution of Gas Engineers.

5. Safe working in sewers and at sewage works. Health and Safety Guideline No 2. 1979. National Joint Health and Safety Committee for the Water Service (NJHSCWS).

6. Guidelines for the selection, maintenance and training in the use of respiratory protective equipment. Health and Safety Guideline No 3. 1979. (NJHSCWS).

7. Entry into confined spaces. Guidance note GS5 1980. Health and Safety Executive.

8. Manual valves — a guide to selection for industrial and commercial gas installations IM/5 1981. British Gas.

9. Code of practice for installation of pipes and meters for town gas. Part 3. Low pressure installation pipes. CP331.1974. British Standards Institute.

10. Standards for non-return valves. IM/14.1980. British Gas.

11. Safety in pressure testing. Guidance note GS4.1977. Health and Safety Executive.

12. Guideance notes on the installation of gas pipework boosters and compressors in customers premises. IM/16. 1982. British Gas.

13. Electrical equipment in gas production, transmission, storage and distribution. Communication 855 1978. IGE/SR/3. The Institution of Gas Engineers.

14. Purging procedures for non-domestic gas installations. IM/2 1975. British Gas.

15. Code of practice for the use of gas in low temperature plant. IM/18. 1982. British Gas.

16. Code of practice for natural gas fuelled spark ignition and dual fuel engines. IM/17. 1981. British Gas.

17. Occupational Exposure Limits. Guidance Note EH40. 1985. Health and Safety Executive.

18. Industrial use of flammable gas detectors. Guidance note CS1. Health and Safety Executive.

19. Code of practice for the selection, installation and maintenance of electrical apparatus for use in potentially explosive atmospheres (other than mining applications or explosive processing and manufacture). BS 5345 1976.

20. Recommendations on Transmission and Distribution Practice. IGE/TD/3. Edition 2: 1983. Communication 1203 — Distribution Mains. Institute of Gas Engineers.

List of acts and regulations (extant at publication)

		(ISBN)
Health and Safety at Work, etc Act 1974	Ch 37	0105437743
The Public Health Act 1936	Ch 49	010850204X
The Energy Act 1983	Ch 25	0105425834
The Control of Pollution Act 1974	Ch 40	0105440744
The Local Government Miscellaneous (Provisions) Act 1976	Ch 57	0105457760
The Gas Act 1972	Ch 60	0105460729
The Oil and Gas Enterprise Act 1982	Ch 23	0105423823
The Gas Safety Regulations 1972	SI 1972 No 1178	0110211782
The Factories Act 1961	Ch 34	0108500276
The Gas Safety (Installation and Use) Regulation 1984	SI 1984 No 1358	0110473582
The Gas Safety (Rights of Entry) Regulations 1983	SI 1983 No 1575	0110375750
The Pipelines Act 1962	Ch 8	0108500985
The Gas Quality Regulations 1983		0110363639
Gas (Metrication) Regulations 1980		0110078519

NOTES FOR GUIDANCE

Consent to supply gas[1] under SS 29 and 29A of Gas Act 1972[2]
See Section 12 of the Oil and Gas Enterprise Act 1982

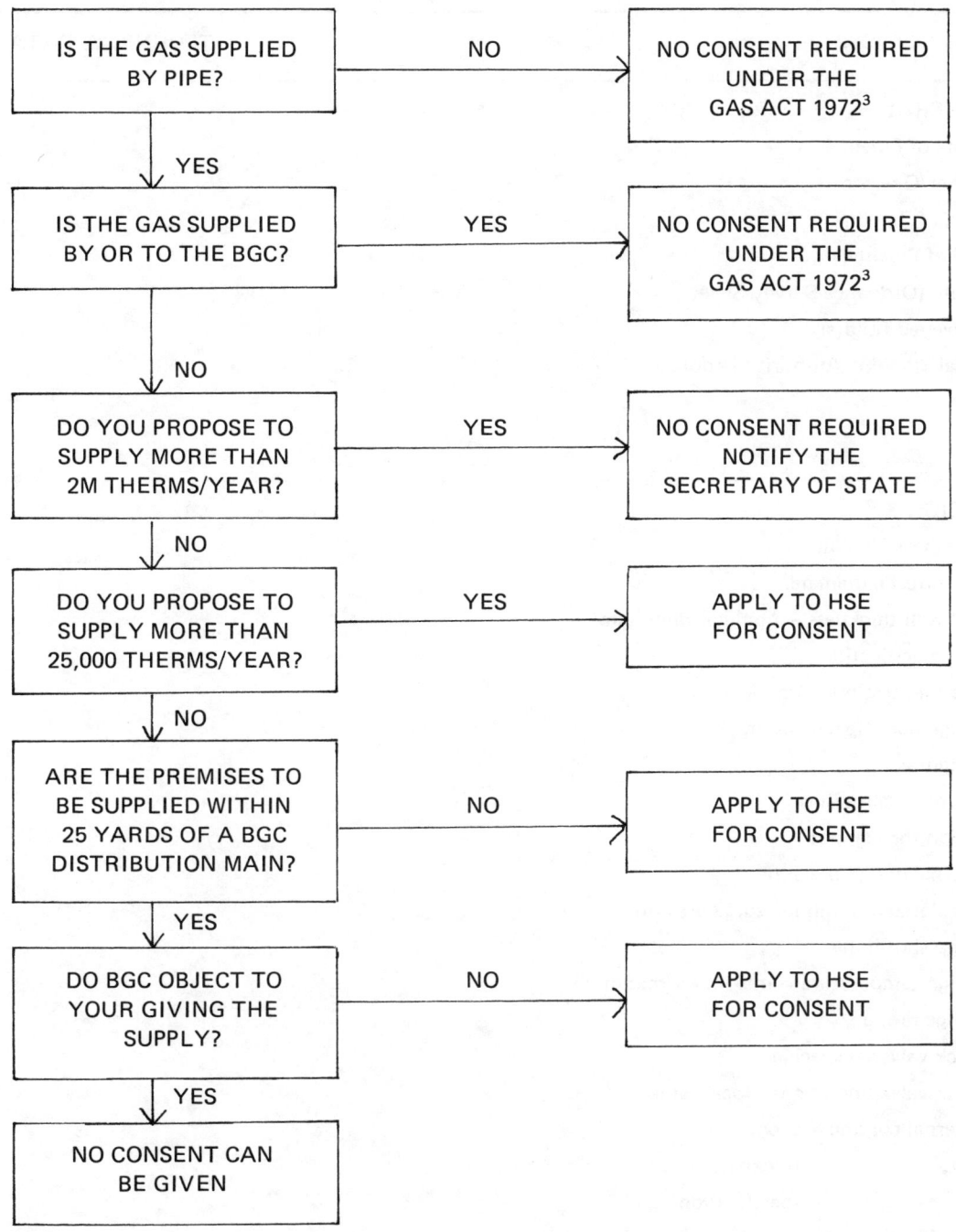

IS THE GAS SUPPLIED BY PIPE? — NO → NO CONSENT REQUIRED UNDER THE GAS ACT 1972[3]

YES ↓

IS THE GAS SUPPLIED BY OR TO THE BGC? — YES → NO CONSENT REQUIRED UNDER THE GAS ACT 1972[3]

NO ↓

DO YOU PROPOSE TO SUPPLY MORE THAN 2M THERMS/YEAR? — YES → NO CONSENT REQUIRED NOTIFY THE SECRETARY OF STATE

NO ↓

DO YOU PROPOSE TO SUPPLY MORE THAN 25,000 THERMS/YEAR? — YES → APPLY TO HSE FOR CONSENT

NO ↓

ARE THE PREMISES TO BE SUPPLIED WITHIN 25 YARDS OF A BGC DISTRIBUTION MAIN? — NO → APPLY TO HSE FOR CONSENT

YES ↓

DO BGC OBJECT TO YOUR GIVING THE SUPPLY? — NO → APPLY TO HSE FOR CONSENT

YES ↓

NO CONSENT CAN BE GIVEN

NOTES

(1) See paragraph 17(b) of Schedule 3 to the Oil and Gas (Enterprise) Act 1982 for the definition of 'gas'.
(2) 'Supply' in this context means 'supply to a single set of premises'.
(3) Other legislation (e.g. Health and Safety at Work etc Act 1974, Pipe-lines Act 1962) may be relevant.
(4) BGC — British Gas Corporation.
(5) HSE — Health and Safety Executive.

Department of Energy

PIPELINES INSPECTORATE

Thames House South

Millbank   London   SW1 4QJ

Form PL 2

PIPE-LINES ACT 1962   TECHNICAL DATA. Sheet 1 of 2

---

1.   GENERAL

1.1   Name of Pipeline

1.2   Owner/Operator

1.3   Start of Pipeline

1.4   End of Pipeline

1.5   Route (Ordnance Survey Sheets)

1.6   Conveyed fluid(s)

1.7   Local Planning Authority (Address)

---

2.   DESIGN

2.1   Length of Pipeline

2.2   Pipe outside diameter

2.3   Pipe wall thickness — nominal/minimum

2.4   Pipe specification

2.5   Pipe mill test pressure

2.6   Design code/specification

2.7   Design factor

2.8   Corrosion allowance

2.9   Design throughput

2.10   Internal design pressure

2.11   Calculated uninhibited surge pressure

2.12   Pump shut-in head

2.13   Design temperature — maximum/minimum

2.14   Flange rating

2.15   Block valves — spacing

2.16   Block valve operation — local/remote

2.17   External coating — type

                        — thickness

                        — specification

2.18   Cathodic Protection — type

                        — test point spacing

                        — specification

2.19   Will permanent pigging facilities by provided?

2.20   Pressure/surge relief system

3.	CONSTRUCTION	
3.1	Construction Code/Specification	
3.2	Depth of cover	
3.3	Welding Code/Specification	
3.4	Radiography — type	
	— percentage	
	— specification	
3.5	Other NDT — type	
	— percentage	
	— specification	
3.6	Coating tests — type	
	— specification	
3.7	Pressure test — pressure	
	— duration	
	— test medium	
	— specification	

4.	OPERATION	
4.1	Control centre location	
4.2	Pressure control — type	
	— maximum pressure	
4.3	Flow measurement, IN — type	
	— accuracy	
4.4	Flow measurement, OUT — type	
	— accuracy	
4.5	Flow measurement, differential — type	
	— accuracy	
4.6	Emergency shut down system	
4.7	Leak detection method	

5.	INSPECTION	
5.1	Route inspection frequency — from the air	
	— from the ground	
5.2	Cathodic Protection — frequency of test readings	

Signed:

Company:

Date:

Summary of blower/compressor types suitable for landfill gas extraction

Blower/compressor type	Typical maximum delivery pressures (psig)	Relative maintenance costs (% capital)	Suitability for LFG
Centrifugal blower:			
Single stage	2	2	Only suited to low pressure extraction
Two stage	5	2	for flaring etc
Rootes type blower:			
Single stage	15	6	Suitable for medium pressure
Two stage	30	6	distribution
Liquid ring pump:			
Single stage	35	5–6	Suitable for wet, dirty gases and
Two stage	100	5–6	relatively high pressure distribution
Sliding vane rotary compressor:			
Single stage	50	6	Suitable for relatively high pressure
Two stage	125	6	distribution
Screw compressor:			
Single stage	60	6	Not generally suitable — too expensive
Two stage	300	6	— unless high pressure required
Reciprocating pump:			
Single stage	50+	10	Expensive when small. Very high
Two stage	125+	10	pressures not usually required
Ejector:			
Single stage	50	4	Suitable for low pressure extraction —
Two stage	100	4	steam or compressed air required

Gas flow metering

Group	Type	Comments	Advantages	Disadvantages
Differential pressure	Orifice plates	These are the simplest and cheapest and the standard design (and others) is outlined in BS 1042	(a) Low cost (b) Can provide (with care) accuracy ± 2% (c) No calibration required if made to a standard specification	(a) Pressure losses (b) Unsuitable for wide variation in flow (c) No use of flows are pulsing (d) Susceptible to dirty fluids or gases
	Nozzles	Available as proprietary items for some flow ranges	(a) High constant value of coefficient of discharge i.e. suitable for wide flow variations (b) Less susceptible to erosion	(a) High pressure loss (b) More expensive than orifice plates
	Pitot tube	Proprietary models available	Can be used for obtaining spot readings of flow rate provided averages of velocity across diameter of pipe are measured i.e. to give flow profiles	Not suitable for long term in-line use where blockages will be likely with dirty gases. Normally used in-line on clean gas with pipe diameters in excess of 400 mm if accuracy required
	Venturi meters	Proprietary models available	Greater accuracy than orifice plates or nozzle. Lower pressure drop	High initial cost. Throat affected by wet or corrosive gases
Variable area	Flow meter	This is the most commonly used and important type of variable flow meter	(a) Wide flow range (b) Unaffected by presence of fittings in line (c) Accuracy comparable to differential pressure types	(a) Unsuitable for high flow rates (b) Larger sizes are expensive (c) Can give problems with very wet gas
Mechanical	(a) Displacement (b) Inferential	Generally the only type of meter acceptable to Department of Trade and Industry under the Gas Act 1972.	(a) Highly accurate (b) Useful for integrated flow record	Extremely expensive
Electronic and ultrasonic	—	Designed for specific uses	—	Extremely expensive

Typical analyses of gases, other than landfill gas

Component % by volume	Gas type				
	Coal mine gas			Natural gas (pipeline)	Digester gas
	Seam (1)	Drainage (2)	Drainage (3)	(4)	(5)
Methane	80—95	75—95	53	95	60—70
Carbon dioxide	0.2—6	0.5—5	1.5	0.3	40—30
Carbon monoxide	0—10	0—10	0.004	—	
Oxygen	—	—	7.9	—	
Nitrogen	2—8	1—25	36	0.6	1
Ethane	8	3	1	4	)
Propane	4	1	0.2	) 0.1	)
Butane	—	1	0.03	)(other HC's)	)1
Helium	)	)	)	—	)
Hydrogen	)0.5	)0.5	)<0.005	—	trace
Hydrogen sulphide	)	)	)	—	0.05—2

Notes

(1)   Taken from D Credy (thesis), Mining Research and Development Establishment.

(2)   Stanhope Bretby, Burton-on-Trent.

(3)   Average values of analyses from two sites in the Western Area of the NCB, representing gas as brought to the surface by a drainage plant and then vented to atmosphere, or sold to customers as industrial fuel gas.

(4)   Taken from "Safety and corrosion in Anaerobic Digestion" by M F Fox. Section 3.6 of "Biomethane, production and uses" Edited by R Buvet, D J Picken ISBN 0-901808-22-9.

(5)   Taken from Engine Division data sheet, Caterpillar Engines, May 1984.

(6)   The values quoted are purely to provide an indication of gas composition at sources as described. Only the most probable sources of gas other than landfill gas, have been identified.

(7)   When carrying out a gas survey in strata, or within buildings, it is likely that some dilution by air will have occurred.

(8)   In addition there may well be preferential adsorption and perhaps degradation of some components. For example methane can be oxidised by aerobic bacteria to carbon dioxide and water, under some conditions and carbon dioxide will be more readily adsorbed by water than methane.

(9)   The ethane/methane ratios in mine gas may vary with time and therefore it is not possible to use ethane as a finger-print to identify the source of gas. Hydrogen and carbon dioxide concentrations are suggested as more readily measurable indicators of the gas source, accepting the reservations made in (8) above.

BIBLIOGRAPHY

Preface

In its deliberations and in the preparation of their report the Landfill Practices Review Group drew upon the practical experience of all sides of the industry as represented by its members and others, commissioned and prepared original papers and inevitably used much published information. Bringing together and summarising existing publications has meant that detailed references and acknowledgement has not been practicable. However, very many valuable publications are available covering the topic of landfilling wastes and the quantity of published literature is growing rapidly.

It would not be practicable to review the whole literature in this report, but it is noted that databases do exist, in particular the comprehensive database compiled, maintained and updated by the

Waste Management Information Bureau
Harwell Laboratory
Oxfordshire OX11 0RA

which contains over 35,000 references covering all aspects of waste management.

The following list of publications and summaries has been compiled as a selection of further relevant reading to indicate to those interested in general or specific aspects of landfilling where further information might be found.

Waste in Landfill Research (WLR)

In addition, the Department of the Environment has published over 60 papers and technical notes as part of the research programme on the Behaviour of Hazardous Wastes in Landfill Sites. Topics include:

— Site investigations (comprising details of the results of investigations of individual landfill sites)

— Pilot scale experiments on leaching from landfills

— Laboratory studies on acid-tars and phenol-bearing lime sludges

— Unsaturated zone studies

— Laboratory column experiments

— Mathematical modelling.

Waste Management Papers (WMP)

Appendix 7.a to Chapter 7 of this report lists the Waste Management Papers published by the Department of the Environment for certain types or groups of wastes, in the form of Technical Memoranda on Arisings, Treatment and Disposal and include Codes of Practice. In addition to those listed in Chapter 7 the following Waste Management Papers have also been published:

WMP No	Title
2	Waste Disposal Surveys
3	Guidelines for the preparation of a waste disposal plan
*4	The licensing of waste disposal sites
5	The relationship between waste disposal authorities and private industry
10	Local Authority Waste Disposal Statistics 1974/75
22	Local Authority Waste Disposal Statistics 1974/75 to 1977/78.

* WMP No 4 will be the subject of review and revision shortly.

The WLR Technical Note series are available from:

Department of the Environment
Publications Sales Unit
Room 4
Building 1
Victoria Road
South Ruislip   HA4 0NZ

Waste Management Papers can be obtained from HMSO.

HOLMES J R (ed) *Practical Waste Management* John Wiley and Sons Ltd (1983) 577pp ISBN 0-471-10491-4 £34.50
Comprehensive book presented covers waste storage and collection, sanitary landfill operations and techniques, leachate and methane generation tipping, licensing, codisposal, plant and equipment, transfer of solid and liquid wastes, reclamation projects, composting baling and incineration. International case studies given.

SUMNER J (Chairman) *Co-operative Programme of Research on the Behaviour of Hazardous Wastes in Landfill Sites* Department of the Environment (1978) Final Report of the Policy Review Committee 169pp ISBN 0-11-751257-5 £7.00
Results of 3 year research programme sponsored by UK Department of Environment presented. Leachate generation, transport and attenuation and influence of site hydrogeological characteristics such as strata permeability are discussed extensively in relation to 20 experimental landfill sites and lysimeter experiments.

COUNTY SURVEYORS' SOCIETY *The Challenge of Waste Disposal* County Surveyors' Society (Jul 1983) 102pp ISBN 0-861-47047-8 £5.00
The UK situation regarding waste disposal is surveyed by examining the legal framework, disposal methods, landfill opportunities and afteruse, waste as a resource, hazardous waste, the role of the private sector and the industrial perspective.

POHLAND F G, ENGLEBRECHT R S *Impact of Sanitary Landfills — An Overview of Environmental Factors and Control Alternatives* American Paper Institute Report, APAPI 1, (Feb 1976) 82pp
Comprehensive review with 147 references of chemistry underlying landfill waste disposal. Particular attention paid to gas and leachate production, attenuation within and outside landfill sites and leachate treatment. Composting and behaviour of microganisms in landfill discussed.

TCHOBANOGLOUS G, THEISEN H, ELIASSEN R *Solid Wastes — Engineering Principles and Management Issues* McGraw-Hill (1977) 621pp ISBN 0-07-063235-9 £16.50
Perspectives, engineering principles and management issues are discussed in detail in an up-to-date review intended for students and practising professionals. Illustrated by case histories, examples of problems. Aspects covered include options in disposal, recovery, storage, transport, handling, collection, transfer, processing, landfill, planning.

SKITT J *Waste Disposal Management and Practice* Charles Knight and Co Ltd, London (1979) 216pp £20.95
Standard reference work, covering refuse composition and analysis, controlled landfill, pulverisation, composting, baling, thermal systems (techniques, gas cleaning, heat utilisation) hazardous wastes, radioactive wastes, planning and design of plant, transfer station, recycling, disposal of sewage sludge with domestic refuse, law, etc.

BEVAN R E *Controlled Tipping of Refuse* Institute of Public Cleansing (1967) 221pp
A classic reference work on chemical processes occurring in landfill, selection and preparation of sites and aspects of tip management. Details of experiments carried out in Manchester (UK) between 1960 and 1965 and many case histories given.

BROMLEY J *The Co-disposal of Toxic Wastes with Domestic and Industrial Waste* AERE Harwell, Environmental and Medical Sciences Division, Didcot, Oxfordshire, AERE-R-9860 (Oct 1980) 25pp Paper from Conf, New Jersey, 3-5 June 1980
Toxic waste disposal reviewed. Co-disposal of cyanide, gasworks, Hg and Cd-bearing, asbestos, PCB, halogenated organic, oily, phenolic and arsenical wastes discussed. Refuse buffered acids and absorbed and retained oils, PCB's etc in limited amounts. Small amounts of cyanide acceptable for co-deposition.

LEHMAN John P (ed) *Hazardous Waste Disposal* Plenum Press New York in cooperation with NATO/CCMS 1983 396pp ISBN 0-306-41171-7
Proceedings of the NATO/CCMS Symposium on Hazardous Waste Disposal held in Washington DC in October 1981 presenting the findings of an eight year joint pilot study on hazardous waste disposal by ten industrialised countries, the OECD, and the EEC, including policy, legislation, regulation, technology and research.

NAYLOR J A, ROWLAND C D, YOUNG C P, BARBER C *The Investigation of Landfill Sites* Water Research Centre, Medmenham Laboratory, PO Box 16, Medmenham, Marlow, Bucks SL7 2HD. TR 91 (Oct 1978) 68pp £3.00
Subjects considered include procedures for site evaluation, techniques of site investigations including drilling and sampling, analytical requirements and long term monitoring. Mechanisms of leachate production and movement within and away from landfills are discussed.

MATRECON INC *Lining of Waste Impoundment and Disposal Facilities* US Environmental Protection Agency, Municipal Environmental Research Laboratory, Cincinnati, Ohio 45268. SW-870 (Sept 1980) 408pp
Topics covered in this useful document include both natural and artificial liners and the reaction of waste liquids with them. The design and operation of lined sites is discussed as are costs and the selection of suitable materials.

LUTTON R J, REGAN G L, JONES L W *Design and Construction of Covers for Solid Waste Landfills* US Environmental Protection Agency, Municipal Environmental Research Laboratory, EPA 600-2-79-165 (Aug 1979) 274pp
Lengthy report which considered mainly various characteristics associated with different soils. These include permeability to liquids and gases, erosion resistance, stability and establishment of vegetation. Use of commercial materials and wastes, minimisation of fire hazard etc also considered. 140 references.

COUNTY SURVEYORS' SOCIETY *Gas Generation from Landfill Sites* Report of the County Surveyors' Society Committee No 4 Special Activity Group 7 Aug 1982 available from Committee Chairman
Report of Working Group covers generation of gas, site examination, detection and measurement of gas, prevention of migration, venting, collection and utilization, guidelines in relation to safety problems and future research. Results of questionnaire to Local Authorities and 17 case studies included.

YOUNG P J and PARKER A *The Identification and Possible Environmental Impact of Trace Gases and Vapours in Landfill Gas* AERE Harwell, Waste Management and Research (1983) 1, pp 213-226

EMCON ASSOCIATES *Methane Generation and Recovery from Landfills* Consolidated Concrete Ltd and Alberta Environment. Ann Arbor Science, Michigan (1980) 147pp ISBN 0-250-40360-9 £8.60
Information presented as following: (1) methane fermentation processes, (2) refuse composition analyses, (3) procedures for estimating theoretical maximum methane yield, (4) models for time-dependency of gas production, (5) gas-flow in landfills, (6) field testing to determine gas recovery potential, (7) gas recovery methods, and (8) gas processing/utilisation.

POHLAND F G *Sanitary Landfill Stabilisation with Leachate Recycle and Residual Treatment* Municipal Environmental Research Laboratory (Oct 1975) (PB 248-524) 112pp
Recirculation of leachate through landfill is highly beneficial in that it increases rate of biological stabilisation and reduces possible environmental contamination. Residual leachate contamination removed by carbon adsorption and/or mixed bed ion exchange.

METRY A, CROSS F L *Leachate Control and Treatment* Technomic Publishing Co Inc, Westport, Connecticut (1976) Volume 7, Environmental Monograph Series 58pp £6.55
Topics covered include leachate generation and characterisation, hydrogeology, leachate migration, control and collection and also treatment. Useful introduction to landfill leachate problems. Some data on landfill odours given.

ROBINSON H D, MARIS P J *Leachate from Domestic Waste — Generation Composition, and Treatment. A Review* Water Research Centre, Medmenham Laboratory, PO Box 16, Medmenham, Marlow, Bucks SL7 2HD Technical Report TR-108 (Mar 1979) 38pp
Results of investigations into generation composition and treatment of leachates reviewed. New data on composition of leachates from domestic waste landfills presented. Best methods of leachate treatment are aerobic/biological, recirculation through landfill and land spraying. 77 references.

BARBER C *Domestic Waste and Leachate* Water Research Centre, Henley Road, Medmenham, Marlow, Bucks SL7 2HD Notes on Water Research No 31 (May 1982) 4pp
Hydrogeological properties of landfill sites (particularly containment sites) and leachate generation/properties reviewed. Methods for on-site management/treatment of leachate discussed especially those studied at UK Water Research Centre, i.e. recirculation through landfill and aerobic biological treatment.

FENN D, COCOZZA E, ISBISTER J, BRAIDS O, YARE B, ROUX P *Procedures manual for Groundwater Monitoring at Solid Waste Disposal Facilities* US Environmental Protection Agency, Solid Waste Management Series, EPA 530-SW-616 (Aug 1977) 283pp
Design of monitoring methods depending on the hydrogeology is discussed. Copious information given on construction of monitoring wells, indicators of leachate con-

tamination, leachate composition, sampling, storage of samples analytical methods and monitoring costs.

ANON *The Engineering Behaviour of Industrial and Urban Fill. Symposium Proceedings, University of Birmingham 23-25 April 1979* The Midland Geotechnical Society c/o Department of Civil Engineering, University of Birmingham, PO Box 363, Birmingham B15 2TT (1979) 448pp £11.15
The main topics covered by this collection of 29 papers are (1) physical, chemical and engineering properties, (2) site investigation, (3) use and treatment of fill and (4) construction on fill and possible settlement.

McLOUGHLIN J (for Environmental Resources Ltd) *The Law and Practice Relating to Pollution Control in the United Kingdom* Graham and Trotman Ltd (for the Commission of European Communities) 20 Fouberts Place, London W1 (1976) 386pp ISBN 0-086010-037-5
An outline of the UK pollution control laws, including the Authorities and the law of pollution control for air, inland waters, coastal waters, seas, deposit of waste on land, noise and vibration, nuclear energy and controls over products.

GARNER J F (Ed), HARRIS D J (Ed) *Control of Pollution Encyclopaedia* Butterworths, London (1977) various paging, ISBN 0-406-20508-6 £33.00
Continuously updated publication listing and annotating relevant sections of United Kingdom legislation on waste on land, pollution of inland waters and the sea, atmospheric pollution and pollution by noise. Also covers powers of courts, inspectors etc.

STIEF K (Ed) *Landfill Lining* (in German) Erich Schmidt (1979) Muell und Abfall Suppl No 15 115pp ISBN 3-503-01768-2 £17.70
A collection of papers covering experience, state of the art and research on lining of landfill sites. Emphasis is on the use of synthetic materials. Various drainage systems are described.

NEUMANN U, VAN DOYEN G *Recultivation of Landfills and Waste Tips* (in German) Erich Schmidt Verlag, Berlin (1979) 70pp ISBN 3-503-01790-9 £14.85
Experience with revegetation of landfill sites is described. Data given for effects on various varieties of trees. Many photographs of sites are included.

KAYSER R (ed) *Gas and Leachate in Landfills (in German) — Gas — und Wasserhaushalt von Muelldeponien* Proceedings of an International symposium, held at the Technical University, Brunswick, W Germany, 29 September-1 October 1982 Vol 33 (1982) 516pp
26 papers cover the generation of methane and leachate production in refuse disposal sites. Planning, decomposition, microbial transformations, lysimeters, co-disposal with sewage, water regimes and budgets, extraction wells external influences gas migration modelling and international experience among topics considered.

AERE HARWELL *Landfill Gas Symposium, Harwell, Oxfordshire, 6 May 1981* Preprints and discussion available from M & SD, Bldg 329, AERE, Harwell, Didcot, Oxon OX11 0RA 106pp £4.00
Contributions included review of factors affects methane production, case histories of landfill gas problems (in UK, West Germany, USA), safety and legal considerations, site reclamation and use of gas as energy source in USA and UK (London Brick Landfill Ltd).

AERE *Landfill Leachate Symposium* 1982, Harwell, Oxon AERE 150pp
Symposium contributions presented cover control of water inflow, absorptive capacity of refuse, landfill management and leachate quality. Leachate control and treatment options. Engineering aspects, treatment plants, leachate attenuation in intergranular aquifiers, water authorities and waste disposal and Pitsea Landfill discussed.

AERE HARWELL *Landfill Completion Symposium, Harwell, Oxfordshire, 25 May 1983* Preprints and discussion available from M & SD Bldg 329, AERE, Harwell, Didcot, Oxon 90+pp £12.000
9 contributions include reviews of landfill design, settlement and leachate problems, soil replacement and land drainage in relation to site completion and restoration. Practical experience at some UK sites. Use of coal wastes as cover material and costs also discussed.

AERE HARWELL, British Anaerobic and Biomass Association *Landfill Gas Workshop* Proceedings of a workshop held at St Catherine's College, Oxford, 28 September 1983. Available from M & SD, Bldg 329, Harwell Laboratory, Didcot, Oxon, OX11 0RA or BABA, The White House, Little Bedwyn, Marlborough, Wilts SN8 3JP 75pp £12.00
8 papers examine resources and economics in the UK, case studies at London Brick Landfill Ltd's Stewartby site and GLC's Aveley site, use for electricity generation, gas cleaning, sale of gas and electricity, safety and legislation and the Department of Energy research programme.

AERE *Landfill Planning Symposium, Harwell, Oxon 4 Sept 1984* Proceedings of the 4th Harwell Waste Management Symposium. Preprints available from M & SD, Bldg 329, AERE, Harwell, Didcot, Oxon OX11 0RA 129pp
Eleven papers presented under 3 broad headings: Waste Management Planning, Planning Legislation and Site Licensing, and Case Studies and Landfill Economics.

THE ECONOMICS OF LANDFILL DISPOSAL (see Chapter 3, paragraph 3.104)

1. Introduction

1.1 The economics of landfill disposal need to be understood by those involved in order that the true cost of development, operation and restoration of a site can be evaluated prior to development; only by so doing can it be demonstrated that the site is financially viable and that its proper restoration can be assured. This Annex aims to provide information on:

(a) the types and order of cost encountered in developing, operating and restoring landfills; and

(b) a method of comparing and assessing costs on a common basis over the lifetime of the site.

The costs presented are used only to indicate the techniques discussed. Their detail and accuracy are only indicative of preliminary site assessment costs and more accurate figures will be needed for those sites subsequently selected for development.

2. Financial assessment of landfills based on unit costs

2.1 The application of costs per unit of waste (unit costs) for each element of site development, operation and restoration can be used to identify the principal costs to be incurred during the development of a site for landfilling wastes. Unit costs can also be used to aid the planning and development of a landfill and to achieve this it is necessary to determine the elements that make up those unit costs for each site.

2.2 Tables have been prepared covering each element of site assessment, development, operation, restoration, and aftercare in terms of cost per appropriate unit (Tables 1-5). All costs are at mid 1984 prices. They have been selected to indicate the range of costs that are likely to be encountered with some allowance for regional variations. The actual calculation will be site specific and should, in the first instance, be based on a site survey undertaken as part of site selection procedures (see Chapter 3, paragraphs 3.106 et seq).

3. Example

3.1 An example of the application of the unit cost assessment approach is given in Tables 7-11. The site illustrated is a typical containment site where surface waters require protection. It is assumed that a leachate collection system will be installed and that the site will be restored for use as public open space. A total capacity for waste disposal of 2 million cubic metres is available and a waste arising of 100,000 tonnes per annum is anticipated. These characteristics are summarised in Table 6.

3.2 Unit cost assessments can be used to give a rapid indication of the effect of rate of waste input in relation to the cost of landfilling. Table 12 summarises the unit costs for landfilling at equivalent sites having widely differing rates of input. For easy comparison the life of each site is kept constant and only the capacity of each site varies. In general, the higher the rate of input the lower the landfill unit costs.

3.3 The effect of variations in the expected waste input, or changes in working practice on the economics of a site are two examples of "sensitivity analysis". This is a technique which enables the effects of the uncertainties inherent in estimated capital and running costs to be assessed. Table 12 and Figure 1 give a breakdown of landfill costs during the life of the example site; by far the most important unit cost is operation (85%), site development (8%) being the second largest cost.

4. Economic appraisal of alternative sites and disposal options

4.1 Capital investment to set up and operate a landfill site does not occur solely in the first year of development. Operating costs are typically viewed as regular expenditures but foreseeable variations in these can occur. For example, changes in the daily waste input rate or working arrangements will influence the amount spent in future years. It is important therefore, when estimating the costs of operating a site to consider the timing of expenditure. Discounted cash flow (DCF) is a technique that can accommodate expenditures arising in different years.

4.2 The DCF technique calculates the "present value" of expenditures expected in the future. This represents the sum of money which would have to be set aside now to finance the landfill development and operation, throughout its life. The present value figure is calculated by "discounting" back the future costs and revenues to their corresponding value in the present year. This is achieved by multiplying the future cost or revenue by the appropriate "discount factor" for the year in which it is expected to arise. Discount factors are a constant set of values for each interest rate and are widely published. The present value figures of costs and revenues in each future year are subsequently summed to derive the "net present value" (NPV). If the NPV is positive then revenues exceed costs and the development is viable at the rate of interest used.

4.3 Only actual costs are included in a DCF calculation and it is important that all costs appear. For example, when a machine is purchased, its cost enters the DCF in that year in full. Depreciation does not enter a DCF. If funds are borrowed, the interest payments and loan repayments should be included. A DCF calculation is given in Table 13 for the example site.

Calculation of unit costs over life of project

4.4 To obtain the cost per tonne in present day prices over the life of the project the net present value (NPV) of the costs needs to be divided by the NPV of the tonneage ie the sum of the annual tonneages discounted. This is the cost per tonne in the sense of the lowest amount which could be charged for the business to remain solvent; it is not necessarily the price which can be obtained. This method is the only way of allowing for tonneages and costs varying over the life of the project.

Working capital

4.5 One important consideration is the working capital required. In the example given in Table 13, it is year 3 before any revenue would be earned but by year 3, some operating costs would be incurred. The working capital should be calculated with due allowance being made for unforeseen expenditure.

Alternative scenarios

4.6 The effects of different daily rates of waste inputs or site lifetimes can be calculated. Tables 14, 15 and 16 consider the example site with the same annual input (100,000 tonnes) but of different capacities and therefore different lifetimes. The annual operating costs, being dependent on the waste input rate, are the same in each case.

Sensitivity analyses

4.7 In the examples given, certain costs and tonneages have been assumed. In practice, there will be considerable uncertainty about many of the figures employed. To understand the effects of this uncertainty, sensitivity analyses should be conducted to obtain an idea of how likely the worst possible outcome is and what the return on the project would be if they occurred, to obtain an idea of the vulnerability of the project in financial terms to changes in costs and tonneages.

5. Cost implications of accepting difficult wastes for landfill

5.1 Further complexities are introduced when codisposal of difficult wastes with household and commercial wastes is proposed. Such sites will require additional works and control measures. Some of these are given in Table 17. Estimated costs are also given and these should be read in conjunction with Tables 1-5 as they are additional to costs incurred at a normal non-difficult waste site.

Notes for Tables 1 to 5

1.   Leachate will be produced during the lifetime of the site. It is assumed for simplicity in these tables that the average rate of leachate generation equals the site area multiplied by 100 mm per year.

2.   It has been assumed that plant will be leased and the operator will be responsible for repairs and maintenance of the plant. However, if the operator chooses to purchase the plant then, as well as including the cost of the compactor in the capital requirements, it would be necessary to allow for the purchase of subsequent items of plant in later years. Plant life will depend on its rate of utilisation and will be approximately:

   compactor — 2 000-2 500 service hours/annum
   bulldozer — 1 800-2 000 service hours/annum
   excavator — 1 500-1 800 service hours/annum
   tractor and brush — 500 hours

3.   The scale of environmental monitoring (Table 3 item 8) will depend on the size of site and also on site specific factors such as proximity to housing, susceptibility to groundwater or surface water systems, etc. The figures quoted are for general guidance only.

TABLE 1 SITE ASSESSMENT COSTS

Item		Costs £ (mid 1984)	Unit	Site quantity	Site costs £'000	Remarks
1	Reconnaissance — including data collection, detailed walk-over survey, preliminary consultation with statutory authority, preparation of report and assessment of site's suitability and problems	3 000— 8 000	per site			
2	Market Survey — Types and quantities of waste available, competition, potential landfill gas users	2 000— 6 000	per site			
3	Preliminary ground investigation:— including say trial pits, shallow auger holes, water sampling					
	Fees	1 000— 4 000	per site			
	Plant hire	1 000— 2 000	Item			
4	Site survey	2 000— 8 000	per site			
5	Full geological and hydrogeological investigation and outline landfill design:—					
	Fees	4 000— 20 000	per site			
	Contract	5 000— 40 000	per site			
6	Planning and Disposal Licence Consultations and applications advertising and Public meetings	1 000— 10 000	per site			
	Public inquiry	15 000—100 000	per site			

TABLE 2   SITE DEVELOPMENT COSTS

Item	Costs £ (mid 1984)	Unit	Site quantity	Site costs £'000	Remarks
1 Acquisition of site: purchase lease or royalty	very variable	per m^3			
2 Surface water interception channel:					
1 m deep	2—8	m			
2 m deep	8—28	m			
3 Groundwater cut-off drain:					
3 m deep	40	m			
5 m deep	50—60	m			
6.5 m deep	70—80	m			
+ well point. Dewatering if required	40	m			
4 Groundwater cut-off up to 6 m deep:					
Slurry cut-off trench	40—60	m^3			
Sheet piled wall	40—60	m^2			
5 Groundwater separation (on base):					
Fabric on broken stone (depends on source of stone)	1—5	m^2			
6 Culvert eg twin 0.6 diameter up to 2 m deep below original ground level under landfill	200	m			
7 Lining:					
Natural, material on site (clay)	0.5—1.00	m^3			
Natural, clay imported	2.00—4.00	m^3			
Natural, wall sealing only (may be operational cost if wall is steep)	3—5+ (cost of clay per m^3 x 3)	m^3 of wall			
Artificial, membrane (HDPE)	4.5—6	m^2			
Artificial, membrane (Butyl rubber or reinforced LDPE)	2.5—3.5	m^2			
Artificial, membrane on wall only	7—8	m^2 of wall			
Artificial, contaminant resistant bentonite (0.3 m thick)	2.5—4	m^2			
8 Leachate collection:					
Herringbone and manholes	2 000	Ha			
Drainage blanket + fabric + manholes	3 500	Ha			
9 Leachate removal:					
Pump	3 000—10 000	Item			
Electricity cable (if required)	6—8	m			
Delivery pipe	8—10	m			
10 Leachate treatment lagoon (civils + M&E)	40—80	m^3 of capacity			
11 Leachate drains, small diameter sewers	10—15	m			
12 Screen bunds: 2.5 m high x 10 m base width	25	m			
13 Topsoil stripping:	0.70—1.20	m^3			
: (if 0.3 m thick	0.20—0.40	m^2)			
14 Earthworks for bunds and road embankments: 1:1.5 or 1:2 side slopes, material on site	1.5—2	m^3			
15 Preparatory excavation for cover material stockpile for cell 1	0.40—1.00	m^3			
16 Access road (embankments included in (4)):					
Tarmac 5.5 m wide	20	m^2			
Hardcore	10	m^2			
(Also allow for compactor travel routes)					

TABLE 2 (continued)

Item		Costs £ (mid 1984)	Unit	Site quantity	Site costs £'000	Remarks
17	Fencing:					
	2 m security, cranked tops incl. gates	8—20	m			
	1 m stockproof	4—6	m			
18	Tree planting:					
	Standard trees	10—15	No.			
	Shrubs	0.50—0.75	No.			
	Hedges	0.60—0.90	m			
19	Site office and base	5 000—30 000	Item			
20	Wheelcleaner	10 000—25 000	Item			
21	Weighbridge (single axle to full)	8 000—25 000	Item			
22	Garages/workshops/fuel stores	4 000—20 000	Item			
23	Gas vents	3 000— 5 000	Ha			
24	Gas utilisation plant	20 000 +	Ha			
25	Site specific requirements: eg					
	Mineshaft slabs	5 000—15 000	No.			
	Gas pipes over slabs	800— 1 200	No.			
	Highway improvement		Item			
	Services diversion		Item			
	Rail sidings	500— 600	m			
	Electricity, water, telephone, sewers etc		Item			
26	Detailed landfill design: 2—10% of capital costs:					
27	Other costs					

SITE DEVELOPMENT COSTS

TABLE 3   SITE OPERATION COSTS

Item		Costs £/yr (mid 1984)	Unit	Site quantity	Site costs £'000	Remarks
1	Wages and salaries	10 000	per employee			
2	Plant lease:					
	Compactor	25 000	Item			
	Bulldozer	10 000	Item			
	Loader	10 000	Item			
	Motor scraper	30 000	Item			
	Tracked backacter	15 000	Item			
	Dump truck	10 000	Item			
	Pumps	1 000	Item			
	Bowser	1 000	Item			
	Mobile lighting	2 000	Item			
	Jet cleaner	1 500	Item			
	Tractor and attachments	4 000	Item			
3	Machine repair and maintenance (per hour of operation):					
	Compactor	5.50	Hour			
	Bulldozer	3.50	Hour			
	Loader	2.50	Hour			
	Motor scraper	8.00	Hour			
	Tracked backacter	3.00	Hour			
	Dump truck	2.50	Hour			
	Pumps	0.25	Hour			
	Bowser	0.10	Hour			
	Mobile lighting	0.25	Hour			
	Jet cleaner	0.25	Hour			
	Tractor and attachments	1.50	Hour			
4	Fuel, oil and lubricants	0.15	per tonne of waste deposited			
5	Imported cover material	0.50–2.00	m^3			
6	Site maintenance (roads, grass cutting, drainage)	5 000–25 000	Site			
7	Environmental control (pests, litter etc)	5 000–20 000	Site			
8	Environmental monitoring	10 000–15 000	Small site			
		15 000–25 000	Medium site			
		25 000–40 000	Large site			
9	Surveying	200– 600	per hectare			
10	Leachate:					
	On-site treatment	0.10–0.50	m^3			
	Disposal to sewer of pre-treated leachate	0.30–0.75	m^3			
	Disposal to sewer of untreated leachate	1.00–3.00	m^3			
11	Gas: Venting	5 000–10 000	per year			
12	Rates	0.10–0.20	per tonne of waste deposited			
		SITE OPERATION COSTS				
13	Site specific overheads including admin, insurance, professional services, office salaries	20% of operating costs	Item			
		TOTAL SITE OPERATION COSTS				

TABLE 4   RESTORATION COSTS

Item	Costs £ (mid 1984)	Unit	Site quantity	Site costs £'000	Remarks
1 Clay capping —					
clay obtained from stockpile or	0.50—1.00	m^3			
virgin clay bought in	2.00—4.00	m^3			
2 Bentonite enhanced capping:					
If 0.3 m enhanced layer	3 (+ cost of soil if not on site)	m^2			
3 Artificial capping membrane:					
Polythene sheet	0.25	m^2			
Low density polyethelene	2.50	m^2			
Sand protection	2.00	m^2			
4 Subsoil replacement from:					
stockpile	0.50—1.00	m^3			
imported source	2.00—4.00	m^3			
5 Topsoil replacement from:					
stockpile	0.60—1.25	m^3			
imported source	2.00—5.00	m^3			
6 Soil improver/conditioner for subsoil/ poor topsoil	300— 500	Ha			
7 Field drainage	1 000—2 000	Ha			
8 Tree planting:					
Standards	10—15	No.			
Whips	0.50—0.75	No.			
Hedges	0.60—0.90	m			
9 Grass cutting, cultivation etc	0.12—0.25	m^2			

TOTAL RESTORATION COSTS

TABLE 5 AFTERCARE COSTS

First 5 years

Item		Costs £/yr (mid 1984)	Unit	Site quantity	Site costs £'000	Remarks
1	Five year maintenance (total)	400—700	Ha			
2	Differential settlement and treatment (total)	150—300	Ha			
3	Leachate in restored phases:					
	On-site treatment	0.10—0.50	m^3			
	Disposal to sewer of pre-treated leachate	0.30—0.75	m^3			
	Disposal to sewer of untreated leachate	1.00—3.00	m^3			
4	Gas venting (total)	300—1 000	Ha			
5	Environmental monitoring (total)	800—2 000	Ha			
	TOTAL AFTERCARE COST IN FIRST 5 YEARS					

Beyond 5 year maintenance period

Item		Costs £/yr (mid 1984)	Unit	Site quantity	Site costs £'000	Remarks
6	Maintenance beyond 5 year period	50	Ha per year			
7	Leachate in restored phases:					
	On-site treatment	0.10—0.50	m^3			
	Disposal to sewer of pre-treated leachate	0.30—0.75	m^3			
	Disposal to sewer of untreated leachate	1.00—3.00	m^3			
8	Gas venting	50—200	Ha per year			
9	Environmental monitoring	100—250	Ha per year			
	AFTERCARE COST BEYOND FIRST 5 YEARS **PER YEAR**					
	Assume 10 years of post 5 year aftercare					

TABLE 6   APPLICATION OF COSTS TO EXAMPLE LANDFILL

Assumed characteristics	Model site
Site geology	Excavated claypit overlying sandstone
Input rate	
tonnes/day*	400
(tonnes/year)	(100 000)
Site capacity (m^3)	2 million
Site life (years)	20
Loads per day	70
Area (Ha)	13.3
Depth (m)	15
Perimeter (m)	1 460
Number of phases (average area of each phase)	20 (0.67 Ha)
Length of surface water channel required (m)	1 095
Length of roads:	
5.5 m wide tarmac (m)	300
5.5 m wide hardcore (m)	600
Number of operatives	6
Number and type of machines[†]	A, C, F, B, 2E, I, J, K
Leachate generation (m^3/year) (for up to 15 years)	13 300
Clay bunds to seal site 3 m wide (m^2)	20 000
Screen bunds required (m)	600
Value of restored landfill per Ha (mid 1984)	£1 000
Restoration scheme	Public open space

*   Assume density of 1 tonne per cubic metre of compacted waste (as placed).

† A	=	Compactor	G	=	Loader
B	=	Dump trunk	H	=	Lighting
C	=	Backactor	I	=	Tractor and attachments
D	=	Scraper	J	=	Jet cleaner
E	=	Pump	K	=	Bowser
F	=	Bulldozer			

TABLE 7 SITE ASSESSMENT COSTS (EXAMPLE SITE)

Item		Costs £ (mid 1984)	Unit	Model site te/ye £'000
1	Reconnaissance — including data collection, detailed walk-over survey, preliminary consultation with statutory authority, and preparation of report and assessment of site's suitability and problems	3 000— 8 000	per site	6
2	Market survey — types and quantities of waste available, competition, potential landfill gas users	2 000— 6 000	per site	5
3	Preliminary ground investigation:— including say trial pits, shallow auger holes, water sampling			
	Fees	1 000— 4 000	per site	3
	Plant hire	1 000— 2 000	Item	2
4	Site survey	2 000— 8 000	per site	6
5	Full geological and hydrogeological SI and outline landfill design:—			
	Fees	4 000— 20 000	per site	5
	SI Contract	5 000— 40 000	per site	10
6	Planning and Disposal Licence consultations and applications and			
	public meetings	1 000— 10 000	per site	6
	public inquiry	15 000—100 000	per site	—

SITE ASSESSMENT COSTS		43
UNIT COST PER TONNE		£0.02

TABLE 8   SITE DEVELOPMENT COSTS (EXAMPLE SITE)

Item		Costs £ (mid 1984)	Unit	Model site 100 000 te/yr £'000
1	Acquisition of site: purchase lease or royalty		m^2	
2	Surface water interception channel:			
	1 m deep	2–8	m	
	2 m deep	8–28	m	16.45
3	Groundwater cut-off drain:			
	3 m deep	40	m	
	5 m deep	50–60	m	
	6.5 m deep	70–80	m	
	+ well point. Dewatering if required	40	m	
4	Groundwater cut-off up to 6 m deep:			
	Slurry cut-off trench	40–60	m^2	
	Sheet piled wall	40–60	m^2	
5	Groundwater separation (on base):			
	Fabric on broken stone	1–5	m^2	
	(depends on source of stone)			
6	Culvert eg twin 0.6 diameter up to 2 m deep below original ground level under landfill	200	m	
7	Lining:			
	Natural, material on site (clay)	0.50–1.00	m^3	20
	Natural, clay imported	2.00–4.00	m^3	
	Natural, wall sealing only (may be operational cost if wall is steep)	3–5 + (cost of clay per m^3 x 3)	m^3 of wall	
	Artificial, membrane (HDPE)	4.5–6	m^2	
	Artificial, membrane (Butyl rubber or reinforced LDPE)	2.5–3.5	m^2	
	Artificial, membrane on wall only	7–8	m^2 of wall	
	Artificial, contaminant resistant bentonite (0.3 m thick)	2.5–4	m^2	
8	Leachate collection:			
	Herringbone and manholes	2 000	Ha	27
	Drainage blanket + fabric + manholes	3 500	Ha	
9	Leachate removal:			
	Pump	3 000–10 000	Item	7
	Electricity cable (if required)	6–8	m	2.5
	Delivery pipe	8–10	m	3.5
10	Leachate treatment lagoon (civils + M&E) 500–1 500 m^3	40–80	m^3 of capacity	60
11	Leachate drains, small diameter sewers	10–15	m	50
12	Screen bunds: 2.5 m high x 10 m base width	25	m	15
13	Topsoil stripping:	0.70–1.20	m^3	3
	: (if 0.3 m thick	0.20–0.40	m^2)	
14	Earthworks for bunds and road embankments: 1:1.5 or 1:2 side slopes, material on site	1.5–2	m^3	30
15	Preparatory excavation for cover material stockpile for cell 1	0.40–1.00	m^3	20
16	Access road (embankments included in (4)):			
	Tarmac 5.5 m wide	20	m^2	33
	Hardcore	10	m^2	33
	(Also allow for compactor travel routes)			

TABLE 8 (continued)

Item		Costs £ (mid 1984)	Unit	Model site 100 000 te/yr £'000
17	Fencing: 2 m security, cranked tops incl. gates 1 m stockprooof	8—20 4—6	m m	21.9
18	Tree planting: Standard trees Shrubs Hedges	10—15 0.50—0.76 0.60—0.90	No. No. m	2.5 2.5 0.7
19	Site office and base	5 000—30 000	Item	20
20	Wheelcleaner	10 000—25 000	Item	15
21	Weighbridge (single axle to full)	8 000—25 000	Item	15
22	Garages/woorkshops/fuel stores	4 000—20 000	Item	10
23	Gas vents	3 000— 5 000	Ha	39.9
24	Gas utilisation	20 000 +	Ha	
25	Site specific requirements: eg Mineshaft slabs Gas pipes over slabs Highway improvement Service diversion Rail sidings	5 000—15 000 800— 1 200 500— 600	No. No. Item Item m	 20 15
26	Detailed landfill design: 2—10% of capital costs:			35

SITE DEVELOPMENT COSTS (excluding acquisition costs)		517.95
UNIT COST PER TONNE		£0.26

TABLE 9 SITE OPERATION COSTS (EXAMPLE SITE)

Item		Costs £/yr (mid 1984)	Unit	Model site 100 000 te/yr £'000
1	Wages and salaries	10 000	Man	60
2	Plant lease:			
	Compactor	25 000	Item	25
	Bulldozer	10 000	Item	10
	Loader	10 000	Item	—
	Motor scraper	30 000	Item	—
	Tracked backacter	15 000	Item	15
	Dump trunk	10 000	Item	10
	Pumps	1 000	Item	2
	Bowser	1 000	Item	1
	Mobile lighting	2 000	Item	—
	Jet cleaner	1 500	Item	1.50
	Tractor and attachments	4 000	Item	4
3	Machine repair and maintenance:			
	Compactor	5.50	Hour	13
	Bulldozer	3.50	Hour	5
	Loader	2.50	Hour	4
	Motor scraper	8.00	Hour	—
	Tracked backacter	3.00	Hour	3
	Dump trunk	2.50	Hour	3.75
	Pumps	0.25	Hour	1
	Bowser	0.10	Hour	0.50
	Mobile lighting	0.25	Hour	—
	Jet cleaner	0.25	Hour	0.50
	Tractor and attachments	1.50	Hour	3.00
4	Fuel, oil and lubricants	0.15	per tonne of waste deposited	15
5	Imported cover material	0.50–2.00	m^3	
6	Site maintenance (roads, grass cutting, drainage)	5 000–25 000	Site	12
7	Environmental control (pests, wind, etc)	5 000–20 000	Site	12
8	Environmental monitoring	10 000–15 000	Small site	
		15 000–25 000	Medium site	20
		25 000–40 000	Large site	
9	Surveying in operational phase	200–600	per hectare	0.80
10	Leachate:			
	On-site treatment	0.10–0.50	m^3	
	Disposal to sewer of pre-treatment leachate	0.30–0.75	m^3	
	Disposal to sewer of untreated leachate	1.00–3.00	m^3	
11	Gas: venting	5 000–10 000	per year	7
12	Rates	0.10–0.20	per tonne of waste deposited	16
		SITE OPERATION COSTS		245.05
13	Overheads including admin, insurance, professional services, office salaries	20% of operating costs	Item	49.01
		TOTAL SITE OPERATION COSTS PER YEAR		294.06
		UNIT COST PER TONNE		£2.94

TABLE 10 RESTORATION COSTS (EXAMPLE SITE)

Item	Costs £/yr (mid 1984)	Unit	Site 100 000 te/yr £'000
1 Clay capping —			50
clay obtained from stockpile or virgin	0.50—1.00	m^3	
clay bought in	2.00—4.00	m^3	
2 Bentonite enhanced capping: if 0.3 m enhanced layer	3 (+ cost of soil if not on site)	m^2	—
3 Artificial capping membrane:			—
Polythene sheet	0.25	m^2	
Low density polyethelene	2.50	m^2	
Sand protection	2.00	m^2	
4 Subsoil replacement from:			40
stockpile	0.5—1.00	m^3	
imported source	2.00—4.00	m^3	
5 Topsoil replacement from:			50
stockpile	0.60—1.25	m^3	
imported source	2.00—5.00	m^3	
6 Soil improver/conditioner for subsoil/poor topsoil	300— 500	Ha	—
7 Field drainage	1 000—2 000	Ha	13.3
8 Tree planting:			
Standards	10— 15	No.	1.0
Whips	0.50—0.75	No.	2.5
Hedges	0.60—0.90	m	0.5
9 Grass seeding, cultivation etc	0.12—0.25	m^2	20.7
	TOTAL RESTORATION COSTS		178.0
	UNIT COST PER TONNE		£0.09

TABLE 11 AFTERCARE COSTS (EXAMPLE SITE)

First 5 years

Item		Costs £/yr (mid 1984)	Unit	Model site 100 000 te/yr £'000
1	Five year maintenance (total)	400—700	Ha	8
2	Differential settlement and treatment (total)	150—300	Ha	2.0
3	Leachate in restored phases:			
	On-site treatment	0.10—0.50	m^3	20
	Disposal to sewer of pre-treated leachate	0.30—0.75	m^3	33
	Disposal to sewer of untreated leachate	1.00—3.00	m^3	—
4	Gas venting (total)	300—1 000	Ha	4
5	Environmental monitoring (total)	800—2 000	Ha	20
		TOTAL AFTERCARE COST IN FIRST 5 YEARS		87.0
		UNIT COST PER TONNE		£0.04

Beyond 5 year maintenance period

Item		Costs £/yr (mid 1984)	Unit	
6	Maintenance beyond 5 year period	50	Ha per year	0.65
7	Leachate in restored phases:			
	On-site treatment	0.10—0.50	m^3	4.0
	Disposal to sewer of pre-treated leachate	0.30—0.75	m^3	4.0
	Disposal to sewer of untreated leachate	1.00—3.00	m^3	—
8	Gas venting	50—200	Ha per year	0.65
9	Environmental monitoring	100—250	Ha per year	2.65
		AFTERCARE COST BEYOND FIRST 5 YEARS **PER YEAR**		11.95
		Assume 10 years of post 5 year aftercare		119.5
		UNIT COST PER TONNE		£0.06

TABLE 12 LANDFILL UNIT COSTS AND MEAN PERCENTAGE COSTS FOR A RANGE OF WASTE INPUT RATES AT THE EXAMPLE SITE (MID 1984)

	SITE A 25 000 te/yr 0.5 M m³ £	SITE B 50 000 te/yr 1.0 M m³ £	EXAMPLE SITE 100 000 te/yr 2.0 M m³ £	SITE C 150 000 te/yr 3.0 M m³ £	SITE D 250 000 te/yr 5.0 M m³ £	MEAN PERCENTAGE LANDFILL COSTS
Site assessment	0.058	0.035	0.021	0.024	0.014	0.77%
Site development	0.61	0.37	0.26	0.21	0.18	8.36%
Operational costs	5.24	3.36	2.94	2.55	2.05	85.55%
Restoration	0.079	0.079	0.079	0.079	0.079	2.29%
Aftercare	0.104	0.104	0.104	0.104	0.104	3.03%
Unit cost/tonne	£6.09	£3.95	£3.40	£2.97	£2.43	100%

Note:
Additional costs including site acquisition, working capital, head office overheads and research and development may have the effect of increasing these unit costs by a factor between 1.3 and 2.0 times.

TABLE 13   EXAMPLE SITE DISCOUNTED CASH FLOW (MID 1984) (£'000)

Year	0	1	2	3	4	5	6	7	8	9	10	11	12	13
Site assessment	22	21												
Site development														
(a) Initial site preparation works			398											
(b) Progressive site development works			6	6	6	6	6	6	6	6	6	6	6	6
Operational costs				294	294	294	294	294	294	294	294	294	294	294
Restoration costs					8.9	8.9	8.9	8.9	8.9	8.9	8.9	8.9	8.9	8.9
Aftercare														
(a) 5 year maintenance period					0.9	1.8	2.6	3.5	4.4	4.4	4.4	4.4	4.4	4.4
(b) Beyond 5 year maintenance period										0.6	1.2	1.8	2.4	3.0
Total annual cost 1	22	21	404	300	310	311	311	312	313	314	314	315	316	316
Discount factor (5%) 2	1.0	.95	.91	.86	.82	.78	.75	.71	.68	.64	.61	.58	.56	.53
Discounted cost (1 x 2)	22	20	368	258	254	243	233	222	212	201	192	183	177	167
3 input tonnes (000)				100	100	100	100	100	100	100	100	100	100	100
Discounted tonnes (000)				86	82	78	75	71	68	64	61	58	56	53

14	15	16	17	18	19	20	21	22	23	24	25	26-37 (Tot)	Total £'000	
													43	Site assessment
														Site development
													398	(a) Initial site preparation works
6	6	6	6	6	6	6	6						120	(b) Progressive site development works
294	294	294	294	294	294	294	294	294					5880	Operational costs
8.9	8.9	8.9	8.9	8.9	8.9	8.9	8.9	8.9	8.9				178	Restoration costs
														Aftercare
4.4	4.4	4.4	4.4	4.4	4.4	4.4	4.4	4.4	4.4	3.5	2.6	2.7	88	(a) 5 year maintenance period
3.6	4.2	4.8	5.4	6.0	6.0	6.0	6.0	6.0	6.0	6.0	6.0	45	120	(b) Beyond 5 year maintenance period
317	317	318	319	319	319	319	319	313	19	10	9	48	6825	Total annual cost 1
.51	.48	.46	.44	.42	.40	.38	.36	.34	.33	.31	.30	.23		Discount factor (5%) 2
162	152	146	140	134	128	121	115	106	6	3	3	11	3978	Discounted cost (1 x 2)
100	100	100	100	100	100	100	100	100						3 input tonnes (000)
51	48	46	44	42	40	38	36	34					1131	Discounted tonnes (000)

$$\text{Discounted cost per tonne} = \frac{\text{PV cost}}{\text{PV tonnes}} = \frac{£3978}{1131} = £3.52$$

TABLE 14 EXAMPLE SITE DISCOUNTED CASH FLOW SENSITIVITY ANALYSIS (5 YEAR LIFE)

Year	0	1	2	3	4	5	6	7	8	9	10	11	12	13
Site assessment	13	16												
Site development														
(a) Initial site preparation works			225											
(b) Progressive site development works				6	6	6	6	6						
Operational costs				294	294	294	294	294						
Restoration costs					8.9	8.9	8.9	8.9	8.9					
Aftercare														
(a) 5 year maintenance period					0.9	1.8	2.6	3.5	4.4	3.5	2.6	1.8	0.9	
(b) Beyond 5 year maintenance period										0.6	1.2	1.8	2.4	3.0
Total annual cost 1	13	16	231	300	310	311	311	306	13.3	4.1	3.8	3.6	3.3	3.0
Discount factor (5%) 2	1	.95	.91	.86	.82	.78	.75	.71	.68	.64	.61	.58	.56	.53
Discounted cost (1 x 2)	13	15	210	258	254	243	233	217	9	3	2	2	2	2
3 input tonnes (000)				100	100	100	100	100						
Discounted tonnes (000)				86	82	78	75	71						

(MID 1984) (£'000)

14	15	16	17	18	19	20	21	22	23	24	25	Total £'000	
												29	Site assessment
													Site development
												225	(a) Initial site preparation works
												30	(b) Progressive site development works
												1470	Operational costs
												44	Restoration costs
													Aftercare
												22	(a) 5 year maintenance period
3.0	3.0	3.0	3.0	3.0	2.4	1.8	1.2	0.6				30	(b) Beyond 5 year maintenance period
3.0	3.0	3.0	3.0	3.0	2.4	1.8	1.2	0.6				1850	Total annual cost 1
.51	.48	.46	.44	.42	.40	.38	.36	.34	.33	.31	.30		Discount factor (5%) 2
2	1	1	1	1	1	1	–	–				1471	Discounted cost (1 x 2)
													3 input tonnes (000)
												392	Discounted tonnes (000)

Cost per tonne $= \dfrac{\text{PV cost}}{\text{PV tonnes}} = \dfrac{£1471}{392} = £3.75$

TABLE 15 EXAMPLE SITE DISCOUNTED CASH FLOW SENSITIVITY ANALYSIS (10 YEAR

Year	0	1	2	3	4	5	6	7	8	9	10	11	12	13
Site assessment	17	18												
Site development														
(a) Initial site preparation works			283											
(b) Progressive site development works			6	6	6	6	6	6	6	6	6	6		
Operational costs				294	294	294	294	294	294	294	294	294	294	
Restoration costs					8.9	8.9	8.9	8.9	8.9	8.9	8.9	8.9	8.9	8.9
Aftercare														
(a) 5 year maintenance period					0.9	1.8	2.6	3.5	4.4	4.4	4.4	4.4	4.4	4.4
(b) Beyond 5 year maintenance period										0.6	1.2	1.8	2.4	3.0
Total annual cost 1	17	18	289	300	310	311	311	312	313	314	314	315	310	16.3
Discount factor (5%) 2	1.0	.95	.91	.86	.82	.78	.75	.71	.68	.64	.61	.58	.56	.53
Discounted cost (1 x 2)	17	17	263	258	253	242	233	221	212	200	192	182	173	8
3 input tonnes (000)				100	100	100	100	100	100	100	100	100	100	
Discounted tonnes (000)				86	82	78	75	71	68	64	61	58	56	

14	15	16	17	18	19	20	21	22	23	24	25	26-27	Total £'000	
													35	Site assessment
														Site development
													283	(a) Initial site preparation works
													60	(b) Progressive site development works
													2940	Operational costs
													89	Restoration costs
														Aftercare
3.5	2.6	1.8	0.9										44	(a) 5 year maintenance period
3.6	4.2	4.8	5.4	6.0	5.4	4.8	4.2	3.6	3.0	2.4	1.8	1.8	60	(b) Beyond 5 year maintenance period
7.1	6.8	6.6	6.3	6.0	5.4	4.8	4.2	3.6	3.0	2.4	1.8	1.8	3510	Total annual cost 1
.51	.48	.46	.44	.42	.40	.38	.36	.34	.33	.31	.30	.28		Discount factor (5%) 2
4	3	3	3	3	2	2	2	1	1	1	1	.5	2497	Discounted cost (1 x 2)
														3 input tonnes (000)
													699	Discounted tonnes (000)

Cost per tonne $= \dfrac{\text{PV cost}}{\text{PV tonnes}} = \dfrac{£2497}{699} = £3.57$

TABLE 16 EXAMPLE SITE DISCOUNTED CASH FLOW SENSITIVITY ANALYSIS (15 YEAR

Year	0	1	2	3	4	5	6	7	8	9	10	11	12	13
Site assessment	20	20												
Site development														
(a) Initial site preparation works			340											
(b) Progressive site development works				6	6	6	6	6	6	6	6	6	6	6
Operational costs				294	294	294	294	294	294	294	294	294	294	294
Restoration costs					8.9	8.9	8.9	8.9	8.9	8.9	8.9	8.9	8.9	8.9
Aftercare														
(a) 5 year maintenance period					0.9	1.8	2.6	3.5	4.4	4.4	4.4	4.4	4.4	4.4
(b) Beyond 5 year maintenance period										0.6	1.2	1.8	2.4	3.0
Total annual cost 1	20	20	346	300	310	311	312	312	313	314	314	315	316	316
Discount factor (5%) 2	1.0	.95	.91	.86	.82	.78	.75	.71	.68	.64	.61	.58	.56	.53
Discounted cost (1 x 2)	20	19	315	258	254	243	233	222	213	201	192	183	177	167
3 input tonnes (000)				100	100	100	100	100	100	100	100	100	100	100
Discounted tonnes (000)				86	82	78	75	71	68	64	61	58	56	53

LIFE) (MID 1984) (£'000)

14	15	16	17	18	19	20	21	22	23	24	25	26-34	Total £'000	
													40	Site assessment
														Site development
													340	(a) Initial site preparation works
6	6	6											90	(b) Progressive site development works
294	294	294	294										4410	Operational costs
8.9	8.9	8.9	8.9	8.9									133	Restoration costs
														Aftercare
4.4	4.4	4.4	4.4	4.4	4.4	3.5	2.6	1.8	0.9				66	(a) 5 year maintenance period
3.6	4.2	4.8	5.4	6.0	6.0	6.0	6.0	6.0	6.0	5.4	4.8	16.8	90	(b) Beyond 5 year maintenance period
317	317	318	313	19.3	10.4	9.5	8.6	7.8	6.9	5.4	4.8	16.8	5172	Total annual cost 1
.51	.48	.46	.44	.42	.40	.38	.36	.34	.33	.31	.30	.28		Discount factor (5%) 2
162	152	146	138	8.1	4.2	3.6	3.1	2.7	2.3	1.7	1.4	3.9	3326	Discounted cost (1 x 2)
100	100	100	100											3 input tonnes (000)
51	48	46	44										941	Discounted tonnes (000)

$$\text{Cost per tonne} = \frac{\text{PV cost}}{\text{PV tonnes}} = \frac{£3326}{941} = £3.53$$

191

TABLE 17   ADDITIONAL COST CONSIDERATIONS WHEN LANDFILLING DIFFICULT WASTES (MID 1984)

Additional items	Notes	HCI increase over general only site
1 Access and roads:	(i) Main access road: little difference	Nil
	(ii) Secondary roads: may be longer (due to 2 x cells) and better surfaced — possibly	50%
	(iii) Tertiary roads: may be longer and better surfaced — possibly	50—100%
2 Inspection area:	If concrete say 50 x 4 m Pressure jetting equipment Drainage and water supply	£ 2,500 £ 500 £ 1,000
3 Emergency storage:	2 No. areas each 50 x 10 m, bunded Containers: say	£ 1,000 £ 5,000
4 Liquids storage tanks:	2 No. tanks each 100 m^3 2 No. pumps and hoses	£10,000 £ 6,000
5 Weighbridge:	Debatable — say allow 1 No. axle weigher as increase	£ 5,000
6 Noticeboards:	More required but little difference in cost	
7 Civic amenity facilities:	No additional cost	
8 Wheelcleaning:	Unlikely to be any additional cost	
9 Buildings:	Additional changing/washroom, first aid, supervisors office — say	£15,000 max
10 Lighting:	Mast lighting to operational areas: allow for 1 No. additional set	£ 2,500
11 Security:	Allow for 500 m of security fence internally, moved every 3 years	£ 2,000/year
12 Litter control:	Allow for additional fencing and clothing — say	£ 1,000/year
13 Logging control:	Assume grid established on each cell, plus survey at each lift — say	£ 1,500/year
14 Restoration:	Generally higher standard: say cost of additional 300 mm of clay capping	£ 3,000/Ha
15 Cell construction:	The Direct costs of cell construction are for the construction of bunds, sumps etc, for which the increased cost is likely to be very small, and may even result in savings if bigger, and therefore fewer, cells are used. Indirect costs arise out of increased cover requirements, leachate treatment and disposal charges and possibly lower in-place densities	See example
16 Saturated and unsaturated zones:	The effect on costs will range from nothing to the cost of constructing a containment site, and is site specific	See example
17 Base and wall seal:	A wide range, to a maximum of the additional costs of providing a double seal, ie approximately £6/m^2	£0—60,000/Ha
18 Leachate management:	The main increase in costs will reflect any increase in the generation rate during the site's life due to increased cell areas. Unit costs are likely to be between 50p and £1.00 per m^3 of leachate	50p—£1.00 per m^3 of leachate
19 Operational management:	The operational aspects of difficult waste landfills are inextricably bound up with the preparatory works, and in considering the additional costs of landfill because of the disposal of difficult wastes, due regard must be given to increased staffing and plant requirements. It is *estimated* for present purposes that either one or two additional operatives, plus additional management time, plus additional equipment such as an excavator for trenching, and road sweeping plant, will be required at an additional annual cost of between £40,000 and £55,000	£40,000 to £55,000

Annex 4: Figure 1 A REPRESENTATION OF THE LANDFILL COSTS THROUGHOUT THE LIFE OF AN EXAMPLE SITE

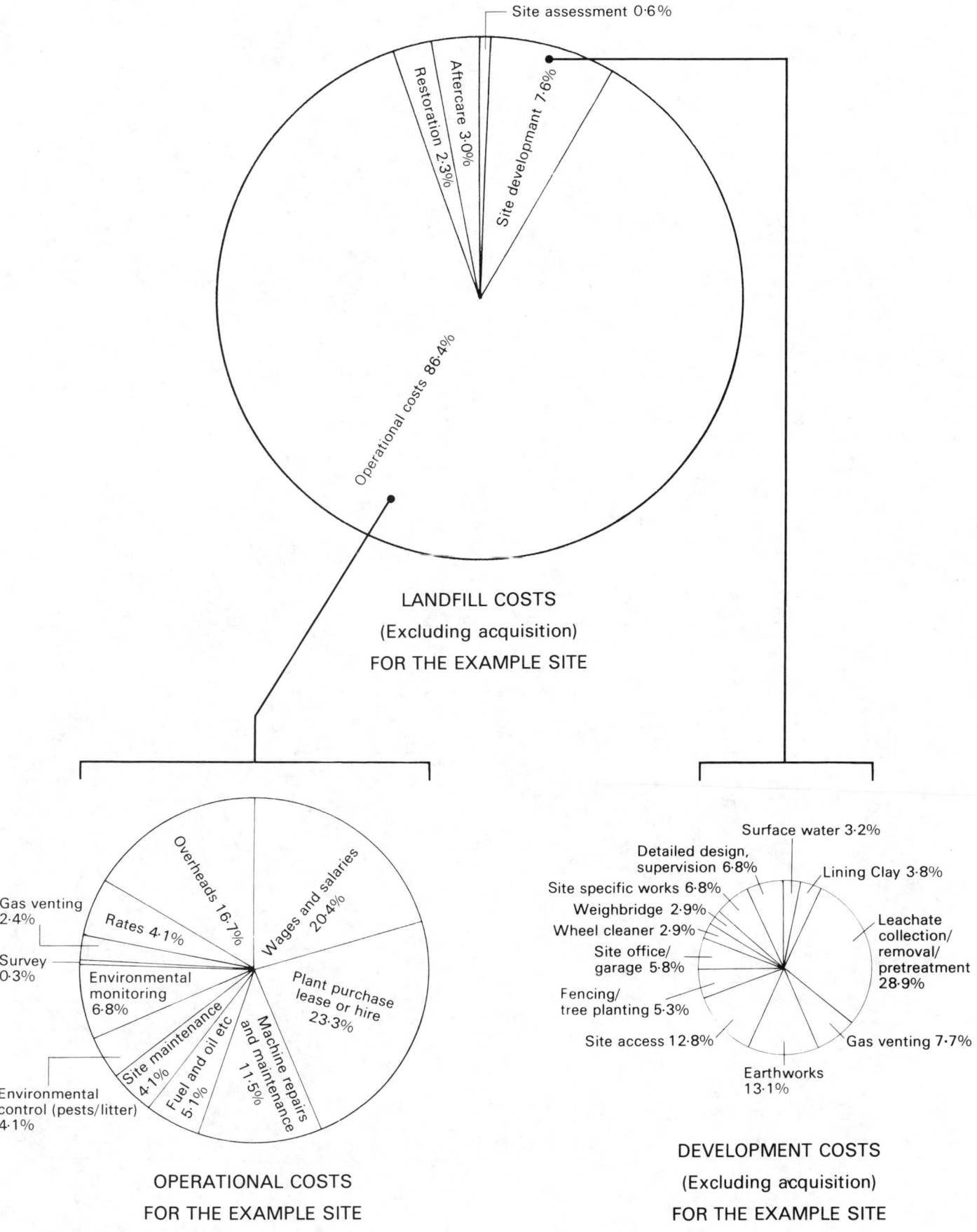

Site assessment 0·6%

Aftercare 3·0%

Restoration 2·3%

Site development 7·6%

Operational costs 86·4%

LANDFILL COSTS

(Excluding acquisition)

FOR THE EXAMPLE SITE

Gas venting 2·4%

Rates 4·1%

Survey 0·3%

Overheads 16·7%

Wages and salaries 20·4%

Environmental monitoring 6·8%

Site maintenance 4·1%

Fuel and oil etc 5·1%

Machine repairs and maintenance 11·5%

Plant purchase lease or hire 23·3%

Environmental control (pests/litter) 4·1%

OPERATIONAL COSTS

FOR THE EXAMPLE SITE

Surface water 3·2%

Detailed design, supervision 6·8%

Lining Clay 3·8%

Site specific works 6·8%

Weighbridge 2·9%

Wheel cleaner 2·9%

Site office/ garage 5·8%

Fencing/ tree planting 5·3%

Site access 12·8%

Leachate collection/ removal/ pretreatment 28·9%

Gas venting 7·7%

Earthworks 13·1%

DEVELOPMENT COSTS

(Excluding acquisition)

FOR THE EXAMPLE SITE

193

GLOSSARY OF TERMS USED IN CONNECTION WITH THE DISPOSAL OF WASTES TO LANDFILL

Preface

The terms used in connection with landfill are derived from many disciplines including chemistry, engineering, physics, geology, microbiology and law. In addition many specialised terms have evolved as landfill practice has developed. It is often difficult to find an adequate definition or explanation of these terms in any one text, particularly in the context of landfill operations as practised in the UK. The terms selected for inclusion in this glossary are strictly related to landfill processes. Mostly, they apply to UK usage but some US terms have also been included. In particular, this has been done where the same term (eg "solid waste") has a different interpretation in each country or where different terms are used to describe the same thing (eg "berm" and "bund"), leading to possible confusion.

'A'

ABSORPTION — The uptake and retention of a liquid or gas into a solid or liquid.

ABSORPTIVE CAPACITY — The maximum amount of liquid taken up and retained by unit weight of solid under specified conditions; usually the amount of liquid retained by unit weight of refuse in a landfill before leachate is produced.

ACID, FATTY — A class of carboxylic acid so called because some members of the class are present in the free form or as glyceryl esters in natural fats and animal or vegetable oils. Fatty acids are compounds of carbon, hydrogen and oxygen that do not contain ring structures. Typical members occurring in landfills include acetic, propionic, butyric, valeric, iso-valeric acids.

ADIT — A horizontal tunnel for access to, or drainage from a mine or similar underground excavation.

ADSORPTION — The uptake and retention of one substance on to the surface of another.

AERATION — Exposure to the action of air.

AEROBIC — See **BACTERIA**.

AEROBIC ZONE — The area within the landfill into which air has penetrated allowing aerobic decomposition of the fill material to proceed.

AFTERCARE — The maintenance work needed to ensure that a restored landfill site does not produce environmental problems.

AFTERUSE — The use to which a landfill site is put following its restoration.

ALIPHATIC — Organic compounds in which the carbon atoms have a straight chain configuration.

ALKALINE — Having a pH value greater than 7.00.

AMMONIACAL NITROGEN — A measure of the nitrogen compounds present — generally from biological activity — by their conversion to ammonia. Concentrations above a few mg/l may indicate pollution.

ANAEROBIC — See **BACTERIA**.

AQUIFER — A subsurface zone or formation of rock which contains exploitable resources of ground water.

Confined aquifer: an aquifer in which the water is confined under pressure by overlying and underlying impermeable strata.

Unconfined aquifer: where the upper surface of a saturated zone forms a water table.

AROMATIC — Organic compounds which contain one or more benzene rings.

ATTENUATE AND DISPERSE — Refers to landfills which allow the migration of leachate and where conversion of the polluting species within the leachate takes place mainly in the strata beneath or around the site.

ATTENUATION — The decrease in concentration of chemical species present in a liquid, eg the decrease in concentration of pollutants in liquid seeping from the foot of a landfill as a result of its passage through the soil.

'B'

BACKACTOR — An excavator mainly used to place soil in landfill restoration and to compact refuse.

BACK BLADING — A levelling technique in which the cutting edge of a tractor-driven earth-mover blade is drawn backwards over waste or cover material.

BACKFILL — The material used for or the act of refilling an excavation.

BACTERIA — Single-cell micro-organisms which multiply by fission. Aerobic bacteria need oxygen for growth. Anaerobic bacteria grow in an oxygen deficient environment. Other bacteria are typified according to the predominant reaction involved eg acetogenic bacteria which break down organic matter to produce acetic acid, methanogenic bacteria form methane from the fatty acids produced by acetogenic and other bacteria.

BALE — To compress solid wastes or recoverable material, using a baling machine or baler, into a block having suitable density and form to allow it to be handled subsequently as a unit. Specially designed high density baling machines can produce bales having a density of 0.9-1.0 t/m^3.

BEARING CAPACITY — The maximum load which the surface of a landfill can support without damage.

BEDROCK — The solid rock underlying soils.

BENTONITE — A type of clay mineral.

BERM (USA) — See **BUND**.

BIOACCUMULATION — Accumulation of (usually) toxic materials within the tissues of living organisms and not readily excreted by them; thus affording their concentration in eg food chains.

BIOCHEMICAL ATTENUATION — The reduction (particularly in leachate) of polluting species by biochemical reaction.

BIOCHEMISTRY — The chemistry of living organisms, including micro-organisms.

BIODEGRADATION — The breakdown of material by the action of micro-organisms.

BIRD STRIKE — Damage caused by birds striking the fuselage or entering the engine of an aircraft.

BLINDING — See **PORE BLINDING**.

BOD (Biochemical Oxygen Demand) — A measure of the amount of material present in water which can be readily oxidised by micro-organisms and is thus a measure of the power of that material to take up the oxygen in water supplies.

BOREHOLE — A hole drilled in the ground or landfill in order to obtain samples of the geological strata, wastes or liquids. Also used as a means of venting or withdrawing gas from landfills (see **MONITORING**).

BORROW MATERIAL — Material extracted from a pit or area adjacent to a landfill for use elsewhere on a landfill site. The pit thereby created may be refilled later using other material.

BUFFER ZONE — (1) A zone within or beneath a landfill where acid or alkaline substances entering that zone can be neutralised by material already present.

(2) An area of land designated to distance landfill sites from adjacent land.

BUFFERING CAPACITY — A measure of the ability of waste to neutralise acidic or alkaline wastes added to it.

BULK DENSITY — The density of a material expressed as the ratio of unit mass to unit volume, including voids.

BULLDOZER — See **DOZER**.

BUND — An embankment usually of clay or other inert material used to prevent the lateral movement of wastes. Synonymous with **BERM** (USA).

'C'

CAPILLARY ZONE — Area above the water table into which water is attracted by capillary action.

CAPPING — The covering of a landfill with impervious material to inhibit penetration by liquids.

CARBOXYLIC ACID — An organic compound containing one or more units of the acid group — COOH. See also **ACID, FATTY**.

CAPACITY — See **ABSORPTIVE, ION EXCHANGE, BEARING** and **BUFFERING CAPACITY**.

CATCHMENT AREA — (1) The area from which solid waste is collected for a specific landfill or transfer station.

(2) The area from which water drains into a reservoir, river or lake or landfill.

CELL — The compartment within a landfill in which waste is deposited. The cell has defined boundaries which may be a low permeability base, a bund wall and low permeability cover.

CELLULOSE — Organic material present in wood, cotton and other fibrous materials.

CHALK PUDDLING — Crushing limestone or chalk and mixing it with water and/or clay and compacting the resultant slurry to line the base of a landfill. The puddled chalk provides a less pervious layer.

CHEMICAL FIXATION — See **SOLIDICATION**.

COD (Chemical Oxygen Demand) — A measure of the total amount of chemically oxidisable material present in liquid.

CHELATION — See **COMPLEX**.

CODISPOSAL — See **LANDFILL**.

CONTROLLED TIPPING — Also controlled landfill. See **LANDFILL**.

COMPACTING — Increasing the density of solid waste in landfills by the repeated passage of heavy machinery over its surface. Also refers to baling machines and stationary compactors for use in compacting solid waste into containers.

COMPACTION RATIO — The ratio of input volume density to final volume density of compacted solid waste. See also **VOLUME REDUCTION** ratio.

COMPATIBLE, COMPATIBILITY — Used to describe wastes which can be mixed or codisposed without adverse effects.

COMPLEX (CHEMICAL) — Chemical species formed by the combination of a metal cation and a neutral or anionic molecule. Chelation is a term used when the molecular structure of the complex contains a ring.

COMPOST — Organic matter decomposed aerobically and used as a fertilizer or soil conditioner.

CONDITIONER (for soil) — Material added to soil to improve its structure and thereby its ability to support vegetation.

'CONCENTRATE AND CONTAIN' — Formerly used to describe the disposal of difficult wastes at containment landfills.

CONTAINMENT SITE — Landfill site where the rate of release of leachate into the environment is extremely low. Polluting components in wastes are retained within such landfills for sufficient time to allow biodegradation and attenuation processes to have occurred; thus preventing the escape of polluting species at unacceptable concentration.

CONTAMINATION — Contamination is the addition, or the result of the addition, or presence of a material or materials to, or in, another substance to such a degree as to render it unfit for its intended purpose.

CONTAMINATION PLUME — The area of contaminated ground water below (and usually down gradient of) a landfill.

CORE — Material obtained when using a hollow drill to produce a borehole.

COVER — Material used to cover solid wastes deposited in landfills. Daily cover is used to cover each lift or layer at the end of each working day to prevent odours, wind-blown litter, insect or rodent infestation, and water ingress. Intermediate cover refers to cover material deposited over wastes at the end of a particular phase of landfilling. Final cover is the layer or layers of material placed on the surface of a landfill during its restoration.

'D'

DEAMINATION — The loss of nitrogen as ammonia (NH_3) from an organic compound.

DECOMPOSITION — Breakdown of matter into more simple molecules. Decomposition may be cause by physical, chemical or micro-biological action.

DENSITY — The mass per unit volume of a substance. In a landfill context, the ratio of the combined mass of fill material and soil cover to their combined volume.

DEWATERING — The removal of water from sludges or pulps by filtering, centrifuging or other means.

DIFFUSION — The process by which molecules and particles migrate and intermingle, the sole energy source being that due to the natural random motion of the molecules.

'DILUTE AND DISPERSE (UK)' — Formerly used to describe landfill sites at which relatively rapid leachate migration could occur.

DIRECT LANDFILL — See **LANDFILL**.

DISPOSAL AUTHORITY (Waste disposal authority) — Disposal authorities were established by the Local Government Act 1972 (for England and Wales), and the Local Government (Scotland) Act 1973 (for Scotland). They consist of:

in England: the County Councils, in shire counties and Borough/District Councils following abolition of the Metropolitan Counties and the Greater London Council except where the Secretary of State establishes a Statutory Authority;

in Wales: the District councils;

in Scotland: the District and Island Councils.

DISPOSAL LICENCE — See **LICENSING**.

DISSOLVED SOLIDS — Also Total Dissolved Solids, TDS. The solid material dissolved in a solvent, usually water; the solid residue remaining after the solvent has been evaporated.

DOMESTIC REFUSE — See **WASTES, DOMESTIC REFUSE.**

DOMING — (1) In a landfill context, doming is the laying of waste and/or cover material (intermediate and final) such that the centre of the covered area is higher than the periphery to assist surface water runoff and thus minimise water ingress.

(2) The water table within a landfill may present a domed configuration as a result of the disposal of large quantities of liquid waste associated with the variable permeability of the landfilled material.

DOZER, BULLDOZER, CRAWLER DOZER — Earthmoving machines fitted with continuous tracks and scraper blades adapted for use on a landfill.

DRAINS — Channels or pipes used to assist the removal of liquids. Agricultural or field drains are constructed using unsocketed, unglazed earthenware or porous concrete pipes laid end to end with open joints or with continuous perforated plastic pipes. French drains are constructed in a similar manner but with the pipes surrounded by filter material, eg gravel, up to the surface.

DRUM — A container or a barrel, usually mild steel or plastic, in which material is stored. Drummed wastes are wastes contained in drums.

DUMP — To throw away, dispose of in an uncontrolled manner. Also an accumulation of wastes abandoned in an uncontrolled manner.

DUST — Fine particles of solid materials ranging in size from 1-75 um diameter (see British Standard 3405) capable of being resuspended in air and settling only slowly under the influence of gravity.

'E'

EARTHWORKS — Engineering work associated with the movement of soils and materials on a landfill.

ECOLOGY — The study of living organisms in relation to their surroundings.

EFFLUENT — The fluid discharged or emitted to the external environment.

ELECTRICAL CONDUCTIVITY — The ease with which an electric current can be transmitted through a substance. Used as an indication of the concentration of certain impurities in water.

EMISSION — A material which is expelled or released to the environment. Usually applied to gaseous or odorous discharges to atmosphere.

ENCAPSULATION — (1) The enclosure of waste within materials having a low permeability.

(2) A process whereby toxic inorganic wastes are incorporated in an inert low permeability matrix.

ENVIRONMENTAL IMPACT — The total effect of any operation on the surrounding environment.

EROSION — The wearing away and removal of weathered land surfaces by natural agents such as rain, running water, wind, temperature changes and bacteria.

EVAPOTRANSPIRATION — The total water transferred to the atmosphere by evaporation from the soil surface and transpiration by plants.

EXCAVATOR — See **DOZER.**

EXOTHERMIC — A chemical or biological reaction which generates heat.

'F'

FATTY ACID — See **ACID, FATTY** and also **CARBOXYLIC ACID.**

FERROUS METALS — A term used to describe iron and its alloys, eg steels. It is also used to describe the general class of metallic materials containing iron, cobalt and nickel as major components.

FIELD CAPACITY — The maximum quantity of water retained by soil or waste after the gravitational water has drained away.

FILL — See **LANDFILL.**

FINAL COVER — See **COVER.**

FIRE, TIP — See **TIP FIRE.**

FISSURE — A cleft or break in soil, rock or landfill cover surface.

FLAMMABLE — Readily set on fire, easily burnt. Synonymous with inflammable, but is the preferred term.

FLASH POINT — The lowest temperature at which a material will provide sufficient vapour to 'flash' or burn momentarily when an ignition source is applied under specified conditions.

FLY TIPPING — The unregulated and hence illegal dumping of waste.

FRAG — The residual fines from a fragmenter plant in which large objects such as scrapped cars, washing machines etc are size reduced and separated for recycling. 'Frag' is often landfilled.

'G'

GARBAGE (USA) — Solid domestic waste, predominantly food wastes. (Swill).

GAS BARRIER — Any device, used to minimise the lateral flow of gas from a landfill site.

GAS, LANDFILL — See LANDFILL GAS.

GEOLOGICAL FORMATION — An assemblage of rocks which have some charcteristics in common, whether of origin, age or composition. Normally now used as a convenient lithological rock unit in a particular area and which can be mapped. A stratigraphic formation is usually implied.

GRADE — The slope imparted to landfill surfaces designed to assist free drainage of surface water and reduce infiltration.

GRADER — A machine used to form a landfill surface slope.

GRADIENT — Slope.

GRAVITATIONAL WATER — Water which moves downwards through soil or a landfill under the influence of gravity.

GREENFIELD SITE — See LANDFILL.

GROUND COVER — Plants grown to prevent or reduce soil erosion.

GROUNDWATER — Water associated with soil or rocks below the ground surface but is usually taken to mean water in the saturated zone.

GROUT, GROUT CURTAIN — A mixture containing cement with sand, PFA or other material used to fill fissures. It can also be used for sealing trenches or boreholes. Grout curtain: a grout filled trench around a landfill site designed to prevent lateral movement of leachate or landfill gases. It can also be formed by the injection of grout into closely spaced boreholes around the perimeter of a landfill site.

'H'

HALOGENATED ORGANIC COMPOUNDS — Organic chemicals containing one or more of the halogens, fluorine, chlorine, bromine or iodine.

HAMMERMILL — A high speed machine in which waste is disintegrated into smaller pieces by fixed or swinging metal hammers.

HARDSTANDING — A concrete or asphalted area on which vehicles or materials can be parked, cleaned or stored.

HAUL DISTANCE — The distance over which wastes or landfill material must be transported either from (1) the last pick-up point of the collection vehicles, or (2) from the transfer station to the landfill.

HEAVY METALS — A term for those ferrous and non-ferrous metals having a density greater than about 4 which possess properties which may be hazardous in the environment. The term usually includes the metals copper, nickel, zinc, chromium, cadmium, mercury, lead, arsenic, and may include selenium and others.

HIGH DENSITY BALING — See BALE.

HOUSEHOLD WASTES — See WASTES, DOMESTIC.

HYDRAULIC CONTINUITY — Occurs when liquid can flow freely and continuously through strata and in relation to landfill, through and around the fill.

HYDROGEOLOGY — The study of water in rocks.

'I'

IGNITION TEMPERATURE — The temperature at which a substance will commence to burn in the presence of air.

IMPERVIOUS — Used to describe materials, natural or synthetic, which have the ability to resist the passage of fluid through them. This property is not absolute, and a cut-off permeability of 10^{-7}–10^{-8} cm/sec for water is often used to describe a landfill liner material as impervious.

INGRESS, WATER — See WATER INGRESS.

INFILTRATION — The entry of water, usually as rain or melted snow, into soil or a landfill.

INFLAMMABLE — See FLAMMABLE.

INPUT — Amount of waste imported into a landfill during a given period of time.

INTERMEDIATE COVER — See COVER.

INTERSTITIAL WATER — See WATER.

ION EXCHANGE — The exchange of ions of like charge between a solid and liquid medium. Ion exchange capacity is the quantity of ions which may be exchanged by unit amount of an ion exchange medium.

'L'

LAGOON — A land area used to contain liquid. Lagoons may be formed in natural or artificially created depressions below surround ground level.

LAND FARMING — The application of wastes or sludges to the land and thereby facilitating their degradation and incorporation into the top layer of soil. Fertilizer is usually added to assist aerobic breakdown.

LANDFILL — The deposit of waste onto and into land in such a way that pollution or harm to the environment is prevented and, through restoration, to provide land which may be used for another purpose.

'Controlled landfill' — is a disposal practice where wastes are deposited in an orderly planned manner at a site licensed under the Control of Pollution Act 1974.

'Monofill' — a landfill site at which only one type of waste is deposited.

'Multifill' — a landfill site which can accept several types of waste.

'Codisposal' — the landfilling of both industrial and household wastes together in such a way that benefit is derived from biodegradation processes to produce relatively non polluting products.

'Direct landfill' — where domestic or other wastes are deposited without pretreatment.

'Onion skin' landfill — in this method waste is deposited in front of the tip face and then spread upwards over the working face in thin layers by steel wheeled compactors to break up and compact the waste.

'Greenfield Site' — a landfill site located on previously undeveloped land.

LANDFILL GAS — A by-product from the digestion by anaerobic bacteria of putrescible matter present in waste deposited on landfill sites. The gas is predominantly methane (65%) together with carbon dioxide (35%) and trace concentrations of a range of vapours and gases.

LEACHATE — Liquid which seeps through a landfill, and by so doing extracts substances from the deposited waste.

Leachate recirculation — the practice of returning leachate to the upper layers of a landfill, from which it has been abstracted, usually by direct spraying on to its surface.

LEACHATE TREATMENT — A process to reduce the polluting potential of leachate. Such processes can include leachate recirculation, spray irrigation over adjacent grassland and biological and physico-chemical processes.

LEACHING — The process of extracting substances from a material by contacting it with a liquid.

LEL (Lower Explosive Limit) — The lowest percentage concentration by volume of a mixture of flammable gas with air which will propagate a flame at $25^{\circ}C$ and atmospheric pressure.

LICENSING — The granting of formal permission for landfill operations at a specified site. The requirement for a licence and the application procedure are set out in Sections 3 and 5 of the Control of Pollution Act 1974 respectively. Licences are issued by waste disposal authorities.

LIFT — A layer of deposited waste and its associated cover material. Its thickness is usually controlled by the height of the working face being operated.

LINER — A natural or synthetic membrane material, used to line the base and sides of a landfill site to prevent leachate seeping into surrounding geological strata.

LITTER — The haphazard distribution of waste on land. At landfill sites this is usually the light, windblown, fraction in household waste such as paper and plastic which escapes before the waste is compacted and covered.

LITTER SCREEN — A moveable screen used on landfill sites to catch litter and prevent its escape from the site. See also **WINDBREAKS**.

LOADER — A dozer fitted with a bucket. See **DOZER**.

LYSIMETER — An experimental construction used to simulate conditions in a landfill or in the soil below a landfill to allow disposal processes to be studied. Lysimeters may consist of specially constructed cells within a landfill or isolated, in situ, blocks of rock or undisturbed columns of soil.

'M'

MEMBRANE — See **LINER**.

METAL, HEAVY — See **HEAVY METALS**.

METALS, NON-FERROUS — See **NON-FERROUS METALS**.

METHANE — CH_4, a colourless, odourless, flammable gas, formed during the anaerobic decomposition of putrescible matter. It forms explosive mixture in the range 5-15% methane in air.

MICROBE MICRO-ORGANISM — Small organisms, usually single cells which normally are only visible under a microscope. They include algae, bacteria and fungi. See also **BACTERIA**.

MILL — A mechanical device used to reduce the size of solid waste to small particles (see **HAMMERMILL, PULVERISE**).

MOISTURE CONTENT — Weight of moisture (usually water) contained in a sample of waste or soil. Usually determined by drying the sample at $105^{\circ}C$ to constant weight.

MONITORING — A continuous or regular periodic check to determine the environmental impact of landfill operations (qv) to ensure compliance with disposal licence conditions and other statutory environmental safety requirements (see Section 9(a) and (b) of the Control of Pollution Act 1974).

MONODISPOSAL — See **LANDFILL**.

MUNICIPAL WASTE — See **WASTES, MUNICIPAL**.

'N'

NON-FERROUS METALS — Metals which do not contain iron.

NUTRIENTS — Materials used by plants and micro-organisms to sustain life.

'O'

ODOUR — The (unpleasant) smell of a material or collection of materials. The characteristic odour of landfill gas is due mainly to alkyl benzenes and limonene, occasionally and additionally associated with esters and organo-sulphur compounds.

ODOUR THRESHOLD — The lowest concentration at which an odour can be detected by the human nose.

ONION-SKIN METHOD — See **LANDFILL**.

ORGANIC (compound) — A substance containing usually two or more carbon atoms in which carbon-carbon atom chains are formed (see **ALIPHATIC, AROMATIC**).

OXIDATION — The loss of electrons by an atom or ion in a chemical reaction. Originally the term simply meant the addition of oxygen.

OXIDATION-REDUCTION POTENTIAL — See **REDOX**.

'P'

PATHOGEN — A micro-organism responsible for disease.

PERCOLATE — The flow of liquid through material by gravitational effects.

PERIPHERAL DRAIN — A drain provided around the boundary of a site.

PERMEABILITY — A measure of the rate at which a fluid will pass through a medium. The coefficient of permeability of a given fluid is an expression of the rate of flow through unit area and thickness under unit differential pressure at a given temperature. Synonymous with hydraulic conductivity when the fluid is water.

PERCHED WATER — See **WATER, PERCHED**.

PERVIOUS — Able to be passed through. See **PERMEABILITY**.

PFA — See **PULVERISED FUEL ASH**.

pH — A measure of the acidity or alkalinity of a liquid acidic = $pH < 7$, or alkaline = $pH > 7$, pH 7 is neutral.

PIEZOMETER — An instrument for measuring hydraulic pressure.

PIEZOMETRIC LEVEL — The level to which water will rise in a borehole penetrating a confined aquifer. The equilibrium between the hydrostatic pressure and atmospheris pressure.

PLUME — A three dimensional envelope that contains all the material (solid, liquid or gas) emitted by a source into its surroundings.

POACHING — Damage to vegetation by the hooves of animals.

POLLUTION, POLLUTANT — The addition of materials or energy to an existing environmental system to the extent that undesirable changes are produced directly or indirectly in that system. A pollutant is a material or type of energy whose introduction into an environmental system leads to pollution.

PORE BLINDING — The blocking by small particles of the interstices or pores in porous rocks.

POROUS — Containing holes or voids.

POROSITY — The ratio of the space occupied by voids or pores to the total volume of the material containing them.

PRECIPITATE — The solid separated from solution by chemical or physical action and deposited as a sediment.

PRECIPITATION — (1) The process of producing a solid from solution by the formation of a precipitate.

(2) Rainfall, snow or sleet.

PROBE — A tube used to collect samples or allow measurements to be made.

PULVERISE — To break solid waste into small pieces. A pulveriser or fragmentiser is a machine used for grinding, shredding or crushing waste or other materials to reduce its volume.

PULVERISED FUEL ASH (PFA) — Ash resulting from the combustion of coal in power stations.

PUTRESCIBLE — Readily able to be decomposed by bacterial action. Offensive odours usually occur as by-products of the decomposition.

PUTRESCIBLE FRACTION — That part of household wastes which will decompose most readily and which often is responsible for offensive odours; commonly due to the decomposition of food and vegetable matter present in the waste.

PHYTOTOXIC — Toxic to plants and thereby inhibiting their growth.

'R'

RECIRCULATION — See **LEACHATE TREATMENT**.

REDOX — A chemical reaction involving oxidation of one chemical and reduction of another.

REDUCTION — A chemical reduction is one in which electrons are added to an atom or ion.

REDUCTION, REFUSE — See **PULVERISE**.

REFUSE, DOMESTIC — See **WASTES, DOMESTIC**.

RESIDUALS — Material left after combustion of wastes.

RESTORATION — Completion of a landfill site to allow planned afteruse.

RUBBISH — See **WASTE**.

RUBBLE — See **WASTES, DEMOLITION**.

RUNOFF — Rain or melted snow which drains from the land surface and in the case of landfill, drains from the surface of the fill.

'S'

SAMPLING — Collecting a portion of a large amount as representative of the whole.

SATURATE — To fill to capacity with water or liquid. Saturation capacity — see **ABSORPTIVE CAPACITY**.

SATURATED ZONE — A geological stratum in which all the void space is filled with water.

SCRAPER — See **DOZER**.

SCREEN — (1) A mesh, supported vertically, used to capture windblown refuse (paper, plastic etc) ie a litter screen.

(2) A mesh or perforated plate used for separating pulverised or shredded refuse into fractions according to particle size.

(3) A mechanical device used to separate medium and larger sized solid material from an effluent prior to further treatment. The separated solids are called 'screenings'.

SEEPAGE — See **LEACHATE**.

SETTLEMENT — The amount by which a landfill surface sinks below its original level due to compaction by its own weight, or that of landfill machinery.

SEWAGE SLUDGE — Sludge resulting from the treatment of raw sewage. It typically contains 70-90% water, prior to dewatering.

SHREDDER — A mechanical device which tears or cuts material into small pieces, used to reduce the size of refuse, scrap metal, paper, card, plastic pieces etc. See also **HAMMERMILL** and **PULVERISE**.

SIFTINGS, RIDDLINGS — Fine ash which falls through an incinerator grate.

SINK POINT — The lowest point to which surface waters flow underground.

SLUDGE — An intimate mixture of solid and liquid.

SMEARING — Mechanical action on wet soil resulting in the formation of a thin compacted layer possessing low permeability.

SOIL — The medium in which plants live and grow and from which through their roots they obtain water and nutrients.

SOIL EROSION — See **EROSION**.

SOIL HORIZON — The interface between layers of soil showing distinct boundaries and having different appearances and properties.

SOIL STRIPPING — The removal of top soil and subsoil either preparatory to further work or for use as cover material.

SOLIDIFICATION — The treatment of liquid slurries and sludges to produce solid products in which toxic ions or elements present in wastes become trapped and thereby immobilised. (See also **ENCAPSULATION**.)

SOLUBILISE — To chemically change substances from an insoluble to a soluble form.

SOLUBILITY — A measure of the ability of a substance to dissolve in a solvent.

SORPTION — The process by which fluid is taken up by a solid or liquid.

SPECIFIC RETENTION — The volume of water retained in rock interstices after free drainage. It is usually expressed as a percentage of total void space.

SPOIL — Materials removed during mining or mineral extraction. When formed into a mound or artificial hill it is known as a spoil heap.

SPRINKLER SYSTEM — Pipes, pumps and spraying devices used for dispersing leachate over a landfill for leachate recirculation or in the distribution of leachate over grassland.

STABILISATION — As applied to landfill this term includes the degradation of organic matter to stable products, and the settlement of the fill to its rest level. The process can take more than 20 years to complete. The term also refers to the use of plants to prevent soil erosion from the surface of a landfill or spoil heap.

SUBGRADE — The protective layer placed beneath a landfill liner.

SUBSIDENCE — The sinking of the landfill surface due to consolidation and filling of underground void space.

SUBSOIL — The less well structured and less biologically active layer below top soil which acts as a reserve of nutrients and water for plant growth in the top soil.

SURCHARGE — (1) To fill a landfill above final contours to allow for subsequent settlement.

(2) A large heap of solid material, eg rubble, which is moved in stages over the surface of a landfill to compact the fill prior to building on it.

SURFACE FLOW — Liquid flowing over the surface of the landfill cover.

SURVEY — Collection of information, relevant to planning of a landfill.

SUSPENDED SOLIDS — Solid material suspended in liquids.

'T'

TINES — The teeth, or prongs, of a fork, plough or harrow. Prongs fitted to a machine used to lift bales of compacted refuse.

TIP — A place where discarded material from mineral extraction processes is deposited.

TIP FIRE — The slow combustion of deposited waste within a landfill. Such underground fires can be extremely difficult to extinguish.

TIPPING, DIRECTION — The direction in which landfilling is to proceed from an existing working face.

TOC — Total Organic Carbon (TOC) is a measure of the amount of elemental and/or combined carbon present in the 'organic chemical' fraction of that material.

TOE — The base of the working face of the landfill.

TOP COVER — See **COVER**.

TOPSOIL — The biologically active surface layer of soil which provides a medium for the cultivation of plants.

TOTTING — The practice of scavenging a landfill to retrieve material and objects having some commercial, usually scrap, value.

TOXIC (TOXICITY) — A substance or material which when taken in produces a detrimental effect on human, animal or plant life.

TRANSMISSIVITY — The rate at which water flows through a vertical section of unit width, extending the full saturated thickness of an aquifer, under unit hydraulic gradient.

TRANSPIRATION — See **EVAPOTRANSPIRATION**.

TRENCH, CUT OFF — A trench filled with impermeable material, to stop the passage of gas or leachate.

TRENCH METHOD — A method by which solid and/or liquid waste is deposited in trenches excavated, usually, into previously deposited waste.

'U'

UNDERGROUND FIRE — See **TIP FIRE**.

UNSATURATED ZONE — The zone of land which lies above a water table in which the pore space in the soil is not saturated with water.

'V'

VENT — Usually refers to a facility provided in a landfill to permit the escape to atmosphere of gases and vapours generated by deposited waste during biodegradation. Perforated pipes, placed laterally or vertically within the landfill, are sometimes used.

VERMIN — Used collectively to describe insects and small wild animals whose habitat is associated with filth, disease and decay.

VOID RATIO — The relationship between the voids or spaces in deposited refuse and consolidated material.

VOLATILE MATTER — Constituents in waste that can readily evaporate at relatively low temperatures.

VOID SPACE — The space existing between and within solids in refuse or soil.

VOLUME REDUCTION — The ratio between the original and final volume of compacted refuse.

'W'

WASHLAND AREA — An area of land which acts as a holding area for river overspill, often within a natural flood plain.

WASTE — 'Waste' is defined in the Control of Pollution Act 1974 Section 30(1) to include:

(a) any substance which constitutes a scrap material or an effluent or other unwanted surplus substance arising from the application of any process; and

(b) any substance or article which requires to be disposed of as being broken, worn out, contaminated or otherwise spoiled,

but does not include a substance which is an explosive within the meaning of the Explosives Act 1875.

WASTE DISPOSAL — The process of getting rid of unwanted, broken, worn out, contaminated or spoiled materials in an orderly, regulated fashion.

WASTE DISPOSAL AUTHORITY — See **DISPOSAL AUTHORITY.**

WASTES, COMMERCIAL — Waste from shops, offices, businesses and places of entertainment. See Control of Pollution Act 1974 Part 1 Section 30(3)(c).

WASTES, CONTROLLED — "Controlled Waste" is waste described as such in the Control of Pollution Act 1974 Part 1. Section 30(1) which defines 'controlled waste' as "household industrial, and commercial waste or any such waste".

WASTES, DEMOLITION — Masonry and rubble wastes arising from the demolition or reconstruction of buildings or other civil engineering structures.

WASTES, DOMESTIC — Waste or refuse that arises from private houses; synonymous with 'household waste'.

WASTES, HAZARDOUS — A waste that, by virtue of its composition, carries the risk of death, injury, or impairment of health to humans or animals, the pollution of waters, or could have an unacceptable environmental impact (qv) if improperly handled, treated or disposed of. The term should not be used for waste that merely contains a hazardous material or materials. It should be used only to describe wastes that contain sufficient of these materials to render the waste as a whole hazardous within the definition given above.

WASTES, HOUSEHOLD — 'Household waste' is defined in the Control of Pollution Act Section 30(3)(a) as consisting "of waste from a private dwelling or residential home or from premises forming part of a university or school or other educational establishment or forming part of a hospital or nursing home".

WASTES, INDUSTRIAL — 'Industrial waste' is defined in the Control of Pollution Act 1974 Section 30(3)(b) as consisting "of waste from any factory within the meaning of the Factories Act 1961 and any premises occupied by a body corporate established by or under any enactment for the purpose of carrying on under national ownership any industry or part of an industry or any undertaking, excluding waste from any mine or quarry". Generally taken to include waste from any industrial undertaking or organisation.

WASTES, INERT — Wastes that do not undergo any significant physical, chemical or biological transformations when deposited in a landfill.

WASTES, MUNICIPAL — Municipal waste is that waste that is collected and disposed of by or on behalf of a local authority. It will generally consist of household waste some commercial waste and waste taken to civic amenity waste collection/disposal sites by the general public. In addition, it may include road and pavement sweepings, gully emptying wastes, and some construction and demolition waste arising from local authority activities.

WASTES, SPECIAL — A particular class of hazardous wastes, so controlled by regulation that prenotification of their transport and deposit is required to be given to statutory authorities.

The procedure to be followed is described in the Control of Pollution (Special Waste) Regulations 1980, issued under Section 17 of the Control of Pollution Act 1974.

WASTES, TOXIC — That class of 'hazardous waste' constituents in which are harmful to a significant degree.

WASTES, TRADE — See **WASTES, COMMERCIAL.**

WATER AUTHORITY — Duly appointed authorities charged (inter alia) with the responsibility properly to manage the nation's water resources, and the duty to protect these resources from pollution. In England and Wales they are the Regional Water Authorities; in Scotland they are the River Purification Boards.

WATER, CAPILLARY — Water, present in land above the water table, which is held between and around soil particles by capillary attraction.

WATER, GRAVITATIONAL — Water which is free to move through the unsaturated zone (qv) of land by gravitational force.

WATER, INGRESS — The infiltration of water into a landfill.

WATER, INTERSTITIAL — Water held within the interstices of rock.

WATER TABLE — The upper surface of a body of groundwater.

WATER, PERCHED — A saturated zone of water prevented from moving down to the water table by a dish shaped impervious layer or occlusion within a permeable soil stratum. Perched water may also be found within waste itself above impermeable waste materials previously deposited or above layers of intermediate cover material such as clay or silt.

WATER, SURFACE — Any natural body of water with a surface open to the atmosphere.

WEIGHBRIDGE — A machine used to weigh large objects such as vehicles. Used to weigh the quantity of waste received at a landfill site.

WHEEL CLEANING — The process by which dirt and mud adhering to the wheels (and maybe the chassis) of vehicles that have travelled over a landfill site is removed, before they gain access to public roads.

WHITE GOODS — A general term used to describe discarded equipment and appliances, usually made from sheet steel, which incorporate a large void space. Typical examples of 'white goods' include refrigerators, freezers, cookers, washing machines. The term has evolved because, in the past, these goods were traditionally white.

WINDBREAKS — A barrier or screen, designed and installed to prevent the spread of windblown litter from a landfill onto adjacent land.

WORKING FACE — The area of a landfill in which waste is currently being deposited.

Printed in the UK for HMSO Dd 0295817 5/92 C11 531/3 12521

'Z'

ZONE, AEROBIC — See **AEROBIC ZONE.**

ZONE, BUFFER — See **BUFFER ZONE.**

ZONE, SATURATED — See **SATURATED ZONE.**

ZONE, UNSATURATED — See **UNSATURATED ZONE.**